Springer Series in Statistics

Advisors:
D. Brillinger, S. Fienberg, J. Gani,
J. Hartigan, K. Krickeberg

Springer Series in Statistics

I. T. Jolliffe

Principal Component Analysis

With 26 Illustrations

Springer-Verlag
New York Berlin Heidelberg Tokyo

I. T. Jolliffe
Mathematical Institute
University of Kent
Canterbury
Kent CT2 7NF
England

AMS Classification: 62H25

Library of Congress Cataloging-in-Publication Data
Jolliffe, I. T.
  Principal component analysis.
  (Springer series in statistics)
  Bibliography: p.
  Includes index.
  1. Principal components analysis. I. Title. II. Series.
QA278.5.J65 1986   519.5′35   85-27882

Typeset by Asco Trade Typesetting Ltd., North Point, Hong Kong.
Printed and bound by R. R. Donnelley & Sons, Harrisonburg, Virginia.
Printed in the United States of America.

9 8 7 6 5 4 3 2 1

ISBN 0-387-96269-7 Springer-Verlag New York Berlin Heidelberg Tokyo
ISBN 3-540-96269-7 Springer-Verlag Berlin Heidelberg New York Tokyo

Preface

Principal component analysis is probably the oldest and best known of the techniques of multivariate analysis. It was first introduced by Pearson (1901), and developed independently by Hotelling (1933). Like many multivariate methods, it was not widely used until the advent of electronic computers, but it is now well entrenched in virtually every statistical computer package.

The central idea of principal component analysis is to reduce the dimensionality of a data set in which there are a large number of interrelated variables, while retaining as much as possible of the variation present in the data set. This reduction is achieved by transforming to a new set of variables, the principal components, which are uncorrelated, and which are ordered so that the first *few* retain most of the variation present in *all* of the original variables. Computation of the principal components reduces to the solution of an eigenvalue-eigenvector problem for a positive-semidefinite symmetric matrix. Thus, the definition and computation of principal components are straightforward but, as will be seen, this apparently simple technique has a wide variety of different applications, as well as a number of different derivations. Any feelings that principal component analysis is a narrow subject should soon be dispelled by the present book; indeed some quite broad topics which are related to principal component analysis receive no more than a brief mention in the final two chapters.

Although the term 'principal component analysis' is in common usage, and is adopted in this book, other terminology may be encountered for the same technique, particularly outside of the statistical literature. For example, the phrase 'empirical orthogonal functions' is common in meteorology, and in other fields the term 'factor analysis' may be used when 'principal component analysis' is meant. References to 'eigenvector analysis ' or 'latent vector analysis' may also camouflage principal component analysis. Finally, some authors refer to principal components analysis rather than principal component analysis. To save space, the abbreviations PCA and PC will be used frequently in the present text.

The book should be useful to readers with a wide variety of backgrounds. Some knowledge of probability and statistics, and of matrix algebra, is necessary, but this knowledge need not be extensive for much of the book. It is expected, however, that most readers will have had some exposure to multivariate analysis in general before specializing to PCA. Many textbooks on multivariate analysis have a chapter or appendix on matrix algebra, e.g. Mardia *et al.* (1979, Appendix A), Morrison (1976, Chapter 2), Press (1972, Chapter 2), and knowledge of a similar amount of matrix algebra will be useful in the present book.

After an introductory chapter which gives a definition and derivation of PCA, together with a brief historical review, there are three main parts to the book. The first part, comprising Chapters 2 and 3, is mainly theoretical and some small parts of it require rather more knowledge of matrix algebra and vector spaces than is typically given in standard texts on multivariate analysis. However, it is not necessary to read all of these chapters in order to understand the second, and largest, part of the book. Readers who are mainly interested in applications could omit the more theoretical sections, although Sections 2.3, 2.4, 3.3, 3.4 and 3.8 are likely to be valuable to most readers; some knowledge of the singular value decomposition which is discussed in Section 3.5 will also be useful in some of the subsequent chapters.

This second part of the book is concerned with the various applications of PCA, and consists of Chapters 4 to 10 inclusive. Several chapters in this part refer to other statistical techniques, in particular from multivariate analysis. Familiarity with at least the basic ideas of multivariate analysis will therefore be useful, although each technique is explained briefly when it is introduced.

The third part, comprising Chapters 11 and 12, is a mixture of theory and potential applications. A number of extensions, generalizations and uses of PCA in special circumstances are outlined. Many of the topics covered in these chapters are relatively new, or outside the mainstream of statistics and, for several, their practical usefulness has yet to be fully explored. For these reasons they are covered much more briefly than the topics in earlier chapters.

The book is completed by an Appendix which contains two sections. The first describes some numerical algorithms for finding PCs, and the second describes the current availability of routines for performing PCA and related analyses in five well-known computer packages.

The coverage of individual chapters is now described in a little more detail. A standard definition and derivation of PCs is given in Chapter 1, but there are a number of alternative definitions and derivations, both geometric and algebraic, which also lead to PCs. In particular the PCs are 'optimal' linear functions of x with respect to several different criteria, and these various optimality criteria are described in Chapter 2. Also included in Chapter

2 are some other mathematical properties of PCs and a discussion of the use of correlation matrices, as opposed to covariance matrices, to derive PCs.

The derivation in Chapter 1, and all of the material of Chapter 2, is in terms of the *population* properties of a random vector **x**. In practice, a *sample* of data is available, from which to estimate PCs, and Chapter 3 discusses the properties of PCs derived from a sample. Many of these properties correspond to population properties but some, for example those based on the singular value decomposition, are defined only for samples. A certain amount of distribution theory for sample PCs has been derived, almost exclusively asymptotic, and a summary of some of these results, together with related inference procedures, is also included in Chapter 3. Most of the technical details are, however, omitted. In PCA, only the first few PCs are conventionally deemed to be useful. However, some of the properties in Chapters 2 and 3, and an example in Chapter 3, show the potential usefulness of the last few, as well as the first few, PCs. Further uses of the last few PCs will be encountered in Chapters 6, 8 and 10. A final section of Chapter 3 discusses how PCs can sometimes be (approximately) deduced, without calculation, from the patterns of the covariance or correlation matrix.

Although the purpose of PCA, namely to reduce the number of variables from p to m ($\ll p$), is simple, the ways in which the PCs can actually be used are quite varied. At the simplest level, if a few uncorrelated variables (the first few PCs) reproduce most of the variation in all of the original variables, and if, further, these variables are interpretable, then the PCs give an alternative, much simpler, description of the data than the original variables. Examples of this use are given in Chapter 4, while subsequent chapters look at more specialized uses of the PCs.

Chapter 5 describes how PCs may be used to look at data graphically, Other graphical representations based on principal co-ordinate analysis, biplots and correspondence analysis, each of which have connections with PCA, are also discussed.

A common question in PCA is how many PCs are needed to account for 'most' of the variation in the original variables. A large number of rules has been proposed to answer this question, and Chapter 6 describes many of them. When PCA replaces a large set of variables by a much smaller set, the smaller set are new variables (the PCs) rather than a subset of the original variables. However, if a subset of the original variables is preferred, then the PCs can also be used to suggest suitable subsets. How this can be done is also discussed in Chapter 6.

In many texts on multivariate analysis, especially those written by non-statisticians, PCA is treated as though it is part of the factor analysis. Similarly, many computer packages give PCA as one of the options in a factor analysis subroutine. Chapter 7 explains that, although factor analysis and PCA have similar aims, they are, in fact, quite different techniques. There are, however, some ways in which PCA can be used in factor analysis and these are briefly described.

The use of PCA to 'orthogonalize' a regression problem, by replacing a set of highly correlated regressor variables by their PCs, is fairly well known. This technique, and several other related ways of using PCs in regression are discussed in Chapter 8.

Principal component analysis is sometimes used as a preliminary to, or in conjunction with, other statistical techniques, the obvious example being in regression, as described in Chapter 8. Chapter 9 discusses the possible uses of PCA in conjunction with three well-known multivariate techniques, namely discriminant analysis, cluster analysis and canonical correlation analysis.

It has been suggested that PCs, especially the last few, can be useful in the detection of outliers in a data set. This idea is discussed in Chapter 10, together with two different, but related, topics. One of these topics is the robust estimation of PCs when it is suspected that outliers may be present in the data, and the other is the evaluation, using influence functions, of which individual observations have the greatest effect on the PCs.

The last two chapters, 11 and 12, are mostly concerned with modifications or generalizations of PCA. The implications for PCA of special types of data are discussed in Chapter 11, with sections on discrete data, non-independent and time series data, compositional data, data from designed experiments, data with group structure, missing data and goodness-of-fit statistics. Most of these topics are covered rather briefly, as are a number of possible generalizations and adaptations of PCA which are described in Chapter 12.

Throughout the monograph various other multivariate techniques are introduced. For example, principal co-ordinate analysis and correspondence analysis appear in Chapter 5, factor analysis in Chapter 7, cluster analysis, discriminant analysis and canonical correlation analysis in Chapter 9, and multivariate analysis of variance in Chapter 11. However, it has not been the intention to give full coverage of multivariate methods or even to cover all those methods which reduce to eigenvalue problems. The various techniques have been introduced only where they are relevant to PCA and its application, and the relatively large number of techniques which have been mentioned is a direct result of the widely varied ways in which PCA can be used.

Throughout the book, a substantial number of examples are given, using data from a wide variety of areas of applications. However, no exercises have been included, since most potential exercises would fall into two narrow categories. One type would ask for proofs or extensions of the theoy given, in particular, in Chapters 2, 3 and 12, and would be exercises mainly in algebra rather than statistics. The second type would require PCAs to be performed and interpreted for various data sets. This is certainly a useful type of exercise, but many readers will find it most fruitful to analyse their own data sets. Furthermore, although the numerous examples given in the book should provide some guidance, there may not be a single 'correct' interpretation of a PCA.

Acknowledgements

My interest in principal component analysis was initiated, nearly 20 years ago, by John Scott, so he is, in one way, responsible for this book being written.

A number of friends and colleagues have commented on earlier drafts of parts of the book, or helped in other ways. I am grateful to Patricia Calder, Chris Folland, Nick Garnham, Tim Hopkins, Byron Jones, Wojtek Krzanowski, Philip North and Barry Vowden for their assistance and encouragement. Particular thanks are due to John Jeffers and Byron Morgan, who each read the entire text of an earlier version of the book, and made many constructive comments which substantially improved the final product. Any remaining errors and omissions are, of course, my responsibility, and I shall be glad to have them brought to my attention.

I have never ceased to be amazed by the patience and efficiency of Mavis Swain, who has expertly typed virtually all of the text, in its various drafts. I am extremely grateful to her, and also to my wife, Jean, who took over my rôle in the household during the last few hectic weeks of preparation. Finally, thanks to Anna, Jean and Nils for help with indexing and proof-reading.

Contents

CHAPTER 1

Introduction

The central idea of principal component analysis (PCA) is to reduce the dimensionality of a data set which consists of a large number of interrelated variables, while retaining as much as possible of the variation present in the data set. This is achieved by transforming to a new set of variables, the principal components (PCs), which are uncorrelated, and which are ordered so that the first *few* retain most of the variation present in *all* of the original variables.

The present introductory chapter is in two parts. In the first, PCA is defined, and what has become the standard derivation of PCs, in terms of eigenvectors of a covariance matrix, is presented. The second part gives a brief historical review of the development of PCA.

1.1. Definition and Derivation of Principal Components

Suppose that $\mathbf{x}$ is a vector of p random variables, and that the variances of the p random variables and the structure of the covariances or correlations between the p variables are of interest. Unless p is small, or the structure is very simple, it will often not be very helpful to simply look at the p variances and all of the $\frac{1}{2}p(p - 1)$ correlations or covariances. An alternative approach is to look for a few ($\ll p$) derived variables which preserve most of the information given by these variances and correlations or covariances.

Although PCA does not ignore covariances and correlations, it concentrates on variances. The first step is to look for a linear function $\boldsymbol{\alpha}_1' \mathbf{x}$ of the elements of $\mathbf{x}$ which has maximum variance, where $\boldsymbol{\alpha}_1$ is a vector of p constants, $\alpha_{11}, \alpha_{12}, \ldots, \alpha_{1p}$, and $'$ denotes transpose, so that

$$\boldsymbol{\alpha}_1' \mathbf{x} = \alpha_{11} x_1 + \alpha_{12} x_2 + \cdots + \alpha_{1p} x_{1p} = \sum_{j=1}^{p} \alpha_{1j} x_j.$$

Next, look for a linear function $\boldsymbol{\alpha}_2' \mathbf{x}$, uncorrelated with $\boldsymbol{\alpha}_1' \mathbf{x}$, which has maximum variance, and so on, so that at the kth stage a linear function $\boldsymbol{\alpha}_k' \mathbf{x}$ is found which has maximum variance subject to being uncorrelated with $\boldsymbol{\alpha}_1' \mathbf{x}$, $\boldsymbol{\alpha}_2' \mathbf{x}, \ldots, \boldsymbol{\alpha}_{r-1}' \mathbf{x}$. The kth derived variable, $\boldsymbol{\alpha}_k' \mathbf{x}$, is the kth PC. Up to p PCs could be found, but it is hoped, in general, that most of the variation in $\mathbf{x}$ will be accounted for by m PCs, where $m \ll p$. The reduction in complexity which can be achieved by transforming to PCs will be demonstrated in several examples later in the book, but it will be useful here to consider the unrealistic, but simple, case where $p = 2$. The advantage of $p = 2$ is, of course, that the data can be plotted exactly in two dimensions.

Figure 1.1 gives a plot of 50 observations on two highly correlated variables x_1, x_2. There is considerable variation in both variables, though rather more in the direction of x_2 than x_1. If we transform to PCs z_1, z_2, we obtain the plot given in Figure 1.2. It is clear that there is greater variation in the direction of z_1 than in either of the original variables, but very little variation

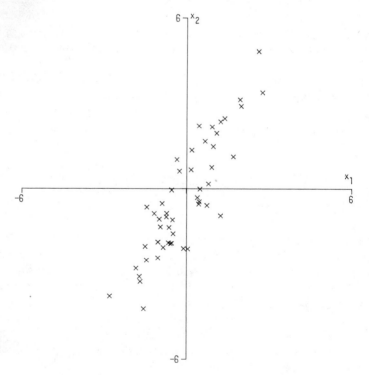

Figure 1.1. Plot of 50 observations on two variables x_1, x_2.

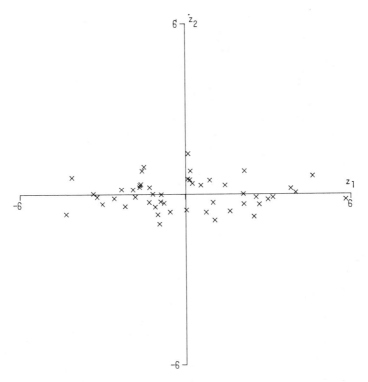

Figure 1.2. Plot of the 50 observations from Figure 1.1 with respect to their PCs z_1, z_2.

in the direction of z_2. More generally, if a set of $p\,(>2)$ variables has substantial correlations among them, then the first few PCs will account for most of the variation in the original variables. Conversely, the last few PCs identify directions in which there is very little variation, i.e. they identify near-constant linear relationships between the original variables.

Having defined PCs, we need to know how to find them. Consider, for the moment, the case where the vector of random variables x has a known covariance matrix, Σ. This is the matrix whose (i, j)th element is the (known) covariance between the ith and jth elements of x when $i \neq j$, and the variance of the jth element of x when $i = j$. (The more realistic case, where Σ is unknown, follows by replacing Σ by a sample covariance matrix S—see Chapter 3.) It turns out that, for $k = 1, 2, \ldots, p$, the kth PC is given by $z_k = \alpha_k' x$ where α_k is an eigenvector of Σ corresponding to its kth largest eigenvalue λ_k. Furthermore, if α_k is chosen to have unit length ($\alpha_k' \alpha_k = 1$), then $\mathrm{var}(z_k) = \lambda_k$, where $\mathrm{var}(z_k)$ denotes the variance of z_k.

The following derivation of the above result is the standard one which is given in many multivariate textbooks, and it may be omitted by readers who mainly are interested in the applications of PCA. Such readers could also skip much of Chapters 2 and 3 and concentrate their attention on Chapters

4–10, although Sections 2.3, 2.4, 3.3, 3.4, 3.8, and to a lesser extent 3.5, are likely to be of interest to most readers.

To derive the form of the PCs, consider first $\alpha_1' x$; α_1 maximizes $\text{var}[\alpha_1' x] = \alpha_1' \Sigma \alpha_1$. It is clear that, as it stands, the maximum will not be achieved for finite α_1, so a normalization constraint must be imposed. The most convenient constraint here is $\alpha_1' \alpha_1 = 1$ (i.e. the sum of squares of elements of α_1 equals 1), but others (e.g. $\text{Max}_j |\alpha_{1j}| = 1$) may more useful in other circumstances, and can easily be substituted later on.

To maximize $\alpha_1' \Sigma \alpha_1$ subject to $\alpha_1' \alpha_1 = 1$, use the technique of Lagrange multipliers and maximize

$$\alpha_1' \Sigma \alpha_1 - \lambda(\alpha_1' \alpha_1 - 1),$$

where λ is a Lagrange multiplier. Differentiation with respect to α_1 gives

$$\Sigma \alpha_1 - \lambda \alpha_1 = \mathbf{0},$$

or

$$(\Sigma - \lambda \mathbf{I}_p)\alpha_1 = \mathbf{0},$$

where $\mathbf{I}_p$ is the $(p \times p)$ identity matrix. Thus, λ is an eigenvalue of Σ and α_1 is the corresponding eigenvector. To decide which of the p eigenvectors is the maximizing value of α_1, note that the quantity to be maximized is

$$\alpha_1' \Sigma \alpha_1 = \alpha_1' \lambda \alpha_1 = \lambda \alpha_1' \alpha_1 = \lambda,$$

so λ must be as large as possible. Thus, α_1 is the eigenvector corresponding to the largest eigenvalue of Σ, and $\text{var}(\alpha_1' x) = \alpha_1' \Sigma \alpha_1 = \lambda_1$, the largest eigenvalue.

In general, the kth PC of x is $\alpha_k' x$ and $\text{var}(\alpha_k' x) = \lambda_k$, where λ_k is the kth largest eigenvalue of Σ, and α_k is the corresponding eigenvector. This will now be proved for $k = 2$; the proof for $k \geq 3$ is slightly more complicated, but very similar.

The second PC, $\alpha_2' x$, maximizes $\alpha_2' \Sigma \alpha_2$ subject to being uncorrelated with $\alpha_1' x$ (i.e. subject to $\text{cov}[\alpha_1' x, \alpha_2' x] = 0$, where $\text{cov}(x, y)$ denotes the covariance between the random variables x and y). But

$$\text{cov}[\alpha_1' x, \alpha_2' x] = \alpha_1' \Sigma \alpha_2 = \alpha_2' \Sigma \alpha_1 = \alpha_2' \lambda_1 \alpha_1' = \lambda_1 \alpha_2' \alpha_1 = \lambda_1 \alpha_1' \alpha_2.$$

Thus any of the equations

$$\alpha_1' \Sigma \alpha_2 = 0, \qquad \alpha_2' \Sigma \alpha_1 = 0,$$

$$\alpha_1' \alpha_2 = 0, \qquad \alpha_2' \alpha_1 = 0,$$

could be used to specify no correlation between $\alpha_1' x$ and $\alpha_1' x$. Choosing the last of these (an arbitrary choice), and noting that, once again, a normalization constraint is necessary, the quantity to be maximized is

$$\alpha_2' \Sigma \alpha_2 - \lambda(\alpha_2' \alpha_2 - 1) - \phi \alpha_2' \alpha_1,$$

where λ, ϕ are Lagrange multipliers. Differentiation with respect to α_2 gives

$$\Sigma\alpha_2 - \lambda\alpha_2 - \phi\alpha_1 = 0$$

and multiplication of this equation on the left by α_1' gives

$$\alpha_1'\Sigma\alpha_2 - \lambda\alpha_1'\alpha_2 - \phi\alpha_1'\alpha_1 = 0,$$

which, since the first two terms are zero and $\alpha_1'\alpha_1 = 1$, reduces to $\phi = 0$. Therefore, $\Sigma\alpha_2 - \lambda\alpha_2 = 0$, i.e. $(\Sigma - \lambda I_p)\alpha_2 = 0$, so, once again, λ is an eigenvalue of Σ and α_2 the corresponding eigenvector.

Again, $\lambda = \alpha_2'\Sigma\alpha_2$, so λ is to be as large as possible. Thus $\lambda = \lambda_2$, the second largest eigenvalue (it cannot equal λ_1 (unless $\lambda_1 = \lambda_2$), since then $\alpha_2 = \alpha_1$ so $\alpha_2'\alpha_1 = 0$ does not hold) and α_2 is the corresponding eigenvector.

As stated above, it can be shown that for the third, fourth, ..., pth PCs, the vectors of coefficients $\alpha_3, \alpha_4, \ldots, \alpha_p$ are the eigenvectors of Σ corresponding to $\lambda_3, \lambda_4, \ldots, \lambda_p$, the third and fourth largest, ..., and the smallest eigenvalue, respectively. Furthermore,

$$\text{var}[\alpha_k'x] = \lambda_k \quad \text{for } k = 1, 2, \ldots, p.$$

1.2. A Brief History of Principal Component Analysis

The earliest descriptions of the technique which is now known as PCA appear to have been given by Pearson (1901) and Hotelling (1933). Hotelling's paper is in two parts and the first, most important, part, together with Pearson's paper, is among the collection of papers edited by Bryant and Atchley (1975).

The two papers adopted different approaches, with the standard algebraic derivation given above being close to that introduced by Hotelling (1933). Pearson (1901), on the other hand, was concerned with finding lines and planes which best fitted a set of points in p-dimensional space, and the geometric optimization problems which he considered lead also to PCs, as will be explained in Section 3.2.

Pearson's comments regarding computation, over 50 years before the wide-spread availability of computers, are interesting. He states that his methods 'can be easily applied to numerical problems', and although he says that the calculations become 'cumbersome' for four or more variables, he suggests that they are still quite feasible.

In the 32 years between Pearson's and Hotelling's papers, very little relevant material seems to have been published, although Rao (1964) indicates that Frisch (1929) adopted a similar approach to that of Pearson. Also, a footnote in Hotelling (1933) suggests that Thurstone (1931) was working along similar lines to Hotelling, but the cited paper, which is also in Bryant and Atchley (1975), is concerned with factor analysis (see Chapter 7), rather than PCA.

Hotelling's approach, too, starts from the ideas of factor analysis but, as will be seen in Chapter 7, PCA, which Hotelling defines, is really rather different in character from factor analysis.

Hotelling's motivation is that there may be a smaller 'fundamental set of independent variables ... which determine the values' of the original p variables. He notes that such variables have been called 'factors' in the psychological literature, but introduces the alternative term 'components' to avoid confusion with other uses of the word 'factor' in mathematics. Hotelling chooses his 'components' so as to maximize their successive contributions to the total of the variances of the original variables, and calls the components which are derived in this way the 'principal components'. The analysis which finds such components is then christened the 'method of principal components'.

Hotelling's derivation of PCs is similar to that given above, using Lagrange multipliers and ending up with an eigenvalue/eigenvector problem, but it differs in three respects. First, he works with a correlation, rather than covariance, matrix (see Section 2.3); second, he looks at the original variables expressed as linear functions of the components rather than components expressed in terms of the original variables; and third, he does not use matrix notation.

After giving the derivation, Hotelling goes on to show how to find the components using the power method (see Appendix A1). He also discusses a different geometric interpretation from that given by Pearson, in terms of ellipsoids of constant probability for multivariate normal distributions (see Section 2.2). A fairly large proportion of his paper, especially the second part, is, however, taken up with material which is not concerned with PCA in its usual form, but rather with factor analysis (see Chapter 7).

A further paper by Hotelling (1936) gave an accelerated version of the power method for finding PCs; in the same year, Girshick (1936) provided some alternative derivations of PCs, and introduced the idea that sample PCs were maximum likelihood estimates of underlying population PCs.

Girshick (1939) also investigated the asymptotic sampling distributions of the coefficients and variances of PCs, but there appears to have been relatively little work on the development of different applications of PCA during the 25 years immediately following publication of Hotelling's paper. Since then, however, an explosion of new applications, and further theoretical developments, has occurred. This expansion reflects the general growth of the statistical literature, but since PCA requires considerable computing power, the expansion of its use coincided with the widespread introduction of electronic computers. Despite Pearson's optimistic comments, it is not really feasible to do PCA by hand, unless p is about four or less. But it is precisely for larger values of p that PCA is most useful, so that the full potential of the technique could not be exploited until after the advent of computers.

To close this section four papers will be mentioned which appeared towards the beginning of the expansion of interest in PCA, and which have become important references within the subject. The first of these, by Anderson (1963), is the most theoretical of the four. It discussed the asymptotic sampling distributions of the coefficients and variances of the sample PCs, building on the earlier work by Girshick (1939), and has been frequently cited in subsequent theoretical developments.

Rao's (1964) paper is remarkable for the large number of new ideas concerning uses, interpretations and extensions of PCA which it introduced.

Gower (1966) discussed some links between PCA and various other statistical techniques, and also provided a number of geometric insights.

Finally, Jeffers (1967) gave an impetus to the really practical side of the subject by discussing two case studies, in which the uses of PCA went beyond that of a simple dimension-reducing tool.

Despite the apparent simplicity of the technique, much research is still being done in the general area of PCA. This is clearly illustrated by the fact that more than 100 of the references in this book appeared in the 1980s. The references also demonstrate the wide variety of areas in which PCA has been applied. Books or articles are cited which include applications in agriculture, biology, chemistry, climatology, demography, ecology, economics, food research, geology, meteorology, psychology and quality control, and it would not be difficult to add further to this list.

CHAPTER 2

Mathematical and Statistical Properties of Population Principal Components

In this chapter many of the mathematical and statistical properties of PCs are described, based on a known population covariance (or correlation) matrix Σ. Further properties are included in Chapter 3 but in the context of sample, rather than population, PCs. As well as being derived from a statistical viewpoint, PCs can be found using purely mathematical arguments; they are given by an orthogonal linear transformation of a set of variables which optimizes a certain algebraic criterion. In fact, the PCs optimize several different algebraic criteria and these optimization properties, together with their statistical implications, are described in the first section of the chapter.

In addition to the algebraic derivation given in Chapter 1, PCs can also be looked at from a geometric viewpoint. The derivation given in the original paper on PCA by Pearson (1901) is geometric but it is relevant to samples, rather than populations, and will therefore be deferred until Section 3.2. However, a number of other properties of population PCs are also geometric in nature and these are discussed in the second section of this chapter.

The first two sections of the chapter concentrate on PCA based on a covariance matrix but the third section describes how a correlation, rather than a covariance, matrix may be used in the derivation of PCs. It also discusses the problems associated with the choice between PCAs based on covariance or correlation matrices.

Throughout most of this text it is assumed that none of the variances of the PCs are equal; nor are they equal to zero. The final section of this chapter explains briefly what happens in the case where there is equality between some of the variances, or when some of the variances are zero.

2.1. Optimal Algebraic Properties of Population Principal Components and Their Statistical Implications

Consider again the derivation of PCs given in Chapter 1, and denote by z the vector whose kth element is z_k, the kth PC, $k = 1, 2, \ldots, p$. (Unless stated otherwise, the kth PC will be taken to mean the PC with the kth largest variance, with corresponding interpretations for the 'kth eigenvalue' and 'kth eigenvector'.) Then

$$z = A'x, \tag{2.1.1}$$

where A is the orthogonal matrix whose kth column, α_k, is the kth eigenvector of Σ. Thus, the PCs are defined by an orthonormal linear transformation of x. Furthermore, we have directly from the derivation in Chapter 1 that

$$\Sigma A = A\Lambda, \tag{2.1.2}$$

where Λ is the diagonal matrix, whose kth diagonal element is λ_k, the kth eigenvalue of Σ, and $\lambda_k = \text{var}(\alpha_k'x) = \text{var}(z_k)$. Two alternative ways of expressing (2.1.2), which follow because A is orthogonal, will be useful later, namely

$$A'\Sigma A = \Lambda, \tag{2.1.3}$$

and

$$\Sigma = A\Lambda A'. \tag{2.1.4}$$

The orthonormal linear transformation of x, (2.1.1), which defines z, has a number of optimal properties which are now discussed in turn.

Property A1. *For any integer q, $1 \le q \le p$, consider the orthonormal linear transformation*

$$y = B'x, \tag{2.1.5}$$

where y is a q-element vector and B' is a $(q \times p)$ matrix, and let $\Sigma_y = B'\Sigma B$ be the variance-covariance matrix for y. Then the trace of Σ_y, denoted $\text{tr}(\Sigma_y)$, is maximized by taking $B = A_q$, where A_q consists of the first q columns of A.

PROOF. Let β_k be the kth column of B; since the columns of A form a basis for p-dimensional space, we have

$$\beta_k = \sum_{j=1}^{p} c_{jk}\alpha_j, \qquad k = 1, 2, \ldots, q,$$

where $c_{jk}, j = 1, 2, \ldots, p, k = 1, 2, \ldots, q$, are appropriately defined constants. Thus $B = AC$ where C is the $(p \times q)$ matrix with (j, k)th element c_{jk}, and

$$\mathbf{B}'\mathbf{\Sigma}\mathbf{B} = \mathbf{C}'\mathbf{A}'\mathbf{\Sigma}\mathbf{A}\mathbf{C} = \mathbf{C}'\mathbf{\Lambda}\mathbf{C}, \quad \text{using (2.1.3)}$$

$$= \sum_{j=1}^{p} \lambda_j \mathbf{c}_j \mathbf{c}_j'$$

where $\mathbf{c}_j'$ is the jth row of $\mathbf{C}$. Therefore

$$\text{tr}(\mathbf{B}'\mathbf{\Sigma}\mathbf{B}) = \sum_{j=1}^{p} \lambda_j \, \text{tr}(\mathbf{c}_j \mathbf{c}_j')$$

$$= \sum_{j=1}^{p} \lambda_j \, \text{tr}(\mathbf{c}_j' \mathbf{c}_j)$$

$$= \sum_{j=1}^{p} \lambda_j \mathbf{c}_j' \mathbf{c}_j$$

$$= \sum_{j=1}^{p} \sum_{k=1}^{q} \lambda_j c_{jk}^2. \tag{2.1.6}$$

Now

$$\mathbf{C} = \mathbf{A}'\mathbf{B}, \quad \text{so}$$

$$\mathbf{C}'\mathbf{C} = \mathbf{B}'\mathbf{A}\mathbf{A}'\mathbf{B} = \mathbf{B}'\mathbf{B} = \mathbf{I}_q,$$

and hence

$$\sum_{j=1}^{p} \sum_{k=1}^{q} c_{jk}^2 = q, \tag{2.1.7}$$

because $\mathbf{A}$ is orthogonal, and the columns of $\mathbf{B}$ are orthonormal. Thus, the columns of $\mathbf{C}$ are also orthonormal, and $\mathbf{C}$ can therefore be thought of as the first q columns of a $(p \times p)$ orthogonal matrix, $\mathbf{D}$, say. But the rows of $\mathbf{D}$ are orthonormal and so satisfy $\mathbf{d}_j'\mathbf{d}_j = 1, j = 1, \ldots, p$. Since the rows of $\mathbf{C}$ consist of the first q elements of the rows of $\mathbf{D}$ it follows that $\mathbf{c}_j'\mathbf{c}_j \leq 1, j = 1, \ldots, p$, i.e.

$$\sum_{k=1}^{q} c_{jk}^2 \leq 1. \tag{2.1.8}$$

Now $\sum_{k=1}^{q} c_{jk}^2$ is the coefficient of λ_j in (2.1.6), the sum of these coefficients is q from (2.1.7), and none of the coefficients can exceed 1, from (2.1.8). Since $\lambda_1 > \lambda_2 > \cdots > \lambda_p$, it is fairly clear that $\sum_{j=1}^{p} (\sum_{k=1}^{q} c_{jk}^2)\lambda_j$ will be maximized if we can find a set of c_{jk}'s for which

$$\sum_{k=1}^{q} c_{jk}^2 = \begin{cases} 1, & j = 1, \ldots, q, \\ 0, & j = q + 1, \ldots, p. \end{cases} \tag{2.1.9}$$

But if $\mathbf{B}' = \mathbf{A}_q'$ then

$$c_{jk} = \begin{cases} 1, & 1 \leq j = k \leq q, \\ 0, & \text{elsewhere}, \end{cases}$$

which satisfies (2.1.9). Thus $\text{tr}(\mathbf{\Sigma}_y)$ achieves its maximum value when $\mathbf{B}' = \mathbf{A}_q'$.

$\square$

Property A2. *Consider again the orthonormal transformation*

$$y = B'x,$$

with x, B, A *and* Σ_y *defined as before. Then* $\mathrm{tr}(\Sigma_y)$ *is minimized by taking* $B = A_q^*$ *where* A_q^* *consists of the last* q *columns of* A.

PROOF. The derivation of PCs given in Chapter 1 can easily be turned around to look, successively, for linear functions of x whose variances are as *small* as possible, subject to being uncorrelated with previous linear functions. The solution is again obtained by finding eigenvectors of Σ, but this time in reverse order, starting with the smallest. The argument which proved Property A1 can similarly be adapted to prove Property A2. $\qquad \square$

The statistical implication of Property A2 is that the last few PCs are not simply unstructured left-overs after removing the important PCs. Because these last PCs have variances as small as possible they are useful in their own right. They can help to detect unsuspected near-constant linear relationships between the elements of x (see Section 3.4), and they may also be useful in regression (Chapter 8), in selecting a subset of variables from x (Section 6.3), and in outlier detection (Section 10.1).

Property A3 (the Spectral Decomposition of Σ).

$$\Sigma = \lambda_1 \alpha_1 \alpha_1' + \lambda_2 \alpha_2 \alpha_2' + \cdots + \lambda_p \alpha_p \alpha_p'. \tag{2.1.10}$$

PROOF.

$$\Sigma = A\Lambda A' \qquad \text{from (2.1.4)},$$

and expansion of the matrix product on the right-hand side of this equation shows it to be equal to

$$\sum_{k=1}^{p} \lambda_k \alpha_k \alpha_k',$$

as required (cf. derivation of (2.1.6)). $\qquad \square$

This result will prove to be useful later. Its statistical implication is that not only can we decompose the combined variances of all the elements of x into decreasing contributions due to each PC, but we can also decompose the whole covariance matrix into contributions $\lambda_k \alpha_k \alpha_k'$ from each PC. Although not strictly decreasing the elements of $\lambda_k \alpha_k \alpha_k'$ will tend to become smaller as k increases, since λ_k decreases for increasing k, whereas the elements of α_k tend to stay 'about the same size' because of the normalization constraints

$$\alpha_k' \alpha_k = 1, \qquad k = 1, 2, \ldots, p.$$

Property A1 emphasizes that the PCs explain, successively, as much as possible of $\mathrm{tr}(\Sigma)$, but the current property shows, intuitively, that they also do a good job of explaining the off-diagonal elements of Σ. This is particularly true when the PCs are derived from a correlation matrix, and is less valid when the covariance matrix is used and the variances of the elements of $\mathbf{x}$ are widely different—see Section 2.3.

Property A4. *As in Properties* A1, A2 *consider the transformation* $\mathbf{y} = \mathbf{B}'\mathbf{x}$. *If* $\det(\Sigma_y)$ *denotes the determinant of the covariance matrix of* $\mathbf{y}$, *then* $\det(\Sigma_y)$ *is maximized when* $\mathbf{B} = \mathbf{A}_q$.

PROOF. Consider any integer, k, between 1 and q, and let $S_k =$ the subspace of p-dimensional vectors orthogonal to $\alpha_1, \ldots, \alpha_{k-1}$. Then $\dim(S_k) = p - k + 1$, where $\dim(S_k)$ denotes the dimension of S_k. The kth eigenvalue, λ_k, of Σ satisfies

$$\lambda_k = \underset{\substack{\alpha \in S_k \\ \alpha \neq 0}}{\text{Sup}} \left\{ \frac{\alpha' \Sigma \alpha}{\alpha' \alpha} \right\}.$$

Suppose that $\mu_1 > \mu_2 > \cdots > \mu_q$ are the eigenvalues of $\mathbf{B}'\Sigma\mathbf{B}$ and that $\gamma_1, \gamma_2, \ldots, \gamma_q$ are the corresponding eigenvectors. Let $T_k =$ the subspace of q-dimensional vectors orthogonal to $\gamma_{k+1}, \ldots, \gamma_q$, with $\dim(T_k) = k$. Then, for any non-zero vector γ in T_k,

$$\frac{\gamma' \mathbf{B}' \Sigma \mathbf{B} \gamma}{\gamma' \gamma} \geq \mu_k.$$

Consider the subspace $\tilde{S}_k$ of p-dimensional vectors of the form $\mathbf{B}\gamma$ for γ in T_k.

$$\dim(\tilde{S}_k) = \dim(T_k) = k \quad \text{(because } \mathbf{B} \text{ is one-to-one, in fact,}$$
$$\mathbf{B} \text{ preserves lengths of vectors).}$$

From a general result concerning dimensions of two vector spaces, we have

$$\dim(S_k \cap \tilde{S}_k) + \dim(S_k + \tilde{S}_k) = \dim S_k + \dim \tilde{S}_k.$$

But

$$\dim(S_k + \tilde{S}_k) \leq p, \quad \dim(S_k) = p - k + 1 \quad \text{and} \quad \dim(\tilde{S}_k) = k,$$

so

$$\dim(S_k \cap \tilde{S}_k) \geq 1.$$

There is therefore a non-zero vector α in S_k of the form $\alpha = \mathbf{B}\gamma$ for a γ in T_k, and it follows that

$$\mu_k \leq \frac{\gamma' \mathbf{B}' \Sigma \mathbf{B} \gamma}{\gamma' \gamma} = \frac{\gamma' \mathbf{B}' \Sigma \mathbf{B} \gamma}{\gamma' \mathbf{B}' \mathbf{B} \gamma} = \frac{\alpha' \Sigma \alpha}{\alpha' \alpha} \leq \lambda_k.$$

Thus the kth eigenvalue of $\mathbf{B}'\Sigma\mathbf{B} \leq k$th eigenvalue of Σ for $k = 1, \ldots, q$. This means that

$$\det(\Sigma_y) = \prod_{k=1}^{q} (k\text{th eigenvalue of } \mathbf{B}'\Sigma\mathbf{B}) \leq \prod_{k=1}^{q} \lambda_k.$$

But if $\mathbf{B} = \mathbf{A}_q$, then the eigenvalues of $\mathbf{B}'\boldsymbol{\Sigma}\mathbf{B}$ are

$$\lambda_1, \lambda_2, \ldots, \lambda_q, \quad \text{so that} \quad \det(\boldsymbol{\Sigma}_y) = \prod_{k=1}^{q} \lambda_k$$

in this case, and therefore $\det(\boldsymbol{\Sigma}_y)$ is maximized when $\mathbf{B} = \mathbf{A}_q$. The result can be extended to the case where the columns of $\mathbf{B}$ are not necessarily ortho-normal, but the diagonal elements of $\mathbf{B}'\mathbf{B}$ are unity—see Okamoto (1969). A stronger, stepwise, version of Property A4 is discussed by O'Hagan (1984), who argues that it provides an alternative derivation of PCs, and that this derivation can be helpful in motivating the use of PCA. O'Hagan's (1984) derivation is, in fact, equivalent to (though a stepwise version of) Property A5, which is discussed next.   □

Note that Property A1 could also have been proved using similar reasoning to that just employed for Property A4, but some of the intermediate results derived during the earlier proof of A1 are useful elsewhere in the chapter.

The statistical importance of the present result follows because the determinant of a covariance matrix, which is called the *generalized variance*, can be used as a single measure of spread for a multivariate random variable (Press, 1972, p. 108). The square root of the generalized variance, for a multivariate normal distribution, is proportional to the 'volume' in p-dimensional space which encloses a fixed proportion of the probability distribution of $\mathbf{x}$. For multivariate normal $\mathbf{x}$, the first q PCs are, therefore, as a consequence of Property A4, q linear functions of $\mathbf{x}$ whose joint probability distribution has contours of fixed probability which enclose the maximum volume.

Property A5. *Suppose that we wish to predict each random variable, x_j, in $\mathbf{x}$ by a linear function of $\mathbf{y}$, where $\mathbf{y} = \mathbf{B}'\mathbf{x}$, as before. If σ_j^2 is the residual variance in predicting x_j from $\mathbf{y}$, then $\sum_{j=1}^{p} \sigma_j^2$ is minimized if $\mathbf{B} = \mathbf{A}_q$.*

The statistical implication of this result is that if we wish to get the best linear predictor of $\mathbf{x}$ (in the sense of minimizing the sum over elements of $\mathbf{x}$ of the residual variances) in a q-dimensional subspace, then this optimal subspace is defined by the first q PCs.

It follows that although Property A5 is stated as an algebraic property, it can equally well be viewed geometrically. In fact, it is essentially the population equivalent of sample Property G3, which is stated and proved in Section 3.2; no proof of the population result, A5, will be given here. Rao (1973, p. 591) outlines a proof in which $\mathbf{y}$ is replaced by an equivalent set of uncorrelated linear functions of $\mathbf{x}$, and it is interesting to note that the PCs are the *only* set of p linear functions of $\mathbf{x}$ which are uncorrelated *and* have orthogonal vectors of coefficients.

A special case of Property A5 was pointed out in Hotelling's (1933) original paper. He notes that the first PC is the linear function of $\mathbf{x}$ which has greater mean square correlation with the elements of $\mathbf{x}$, than does any other linear function.

A modification of Property A5 can be introduced by noting that if $\mathbf{x}$ is predicted by a linear function of $\mathbf{y} = \mathbf{B'x}$, then it follows from standard results from multivariate regression (see, for example, Mardia *et al.*, 1979, p. 160), that the residual covariance matrix for the best such predictor is

$$\mathbf{\Sigma}_x - \mathbf{\Sigma}_{xy}\mathbf{\Sigma}_y^{-1}\mathbf{\Sigma}_{yx}, \tag{2.1.11}$$

where $\mathbf{\Sigma}_x = \mathbf{\Sigma}$, $\mathbf{\Sigma}_y = \mathbf{B'\Sigma B}$, as defined before, $\mathbf{\Sigma}_{xy}$ is the matrix whose (j, k)th element is the covariance between the jth element of $\mathbf{x}$ and the kth element of $\mathbf{y}$, and $\mathbf{\Sigma}_{yx}$ is the transpose of $\mathbf{\Sigma}_{xy}$. Now $\mathbf{\Sigma}_{yx} = \mathbf{B'\Sigma}$, and $\mathbf{\Sigma}_{xy} = \mathbf{\Sigma B}$, so (2.1.11) becomes

$$\mathbf{\Sigma} - \mathbf{\Sigma B}(\mathbf{B'\Sigma B})^{-1}\mathbf{B'\Sigma}. \tag{2.1.12}$$

The diagonal elements of (2.1.12) are $\sigma_j^2, j = 1, 2, \ldots, p$, so, from Property A5, $\mathbf{B} = \mathbf{A}_q$ minimizes

$$\sum_{j=1}^{p} \sigma_j^2 = \text{tr}[\mathbf{\Sigma} - \mathbf{\Sigma B}(\mathbf{B'\Sigma B})^{-1}\mathbf{B'\Sigma}].$$

An alternative criterion is $\|\mathbf{\Sigma} - \mathbf{\Sigma B}(\mathbf{B'\Sigma B})^{-1}\mathbf{B'\Sigma}\|$, where $\|\cdot\|$ denotes the Euclidean norm of a matrix and equals the square root of the sum of squares of *all* the elements in the matrix, and it can be shown (Rao, 1964) that this alternative criterion is also minimized when $\mathbf{B} = \mathbf{A}_q$.

2.2. Geometric Properties of Population Principal Components

It was noted above that Property A5 can be interpreted geometrically, as well as algebraically, and the discussion following Property A4 shows that A4, too, has a geometric interpretation. We now look at two further, purely geometric, properties.

Property G1. *Consider the family of p-dimensional ellipsoids*

$$\mathbf{x'\Sigma}^{-1}\mathbf{x} = \text{const.} \tag{2.2.1}$$

The PCs define the principal axes of these ellipsoids.

PROOF. The PCs are defined by the transformation (2.1.1), $\mathbf{z} = \mathbf{A'x}$, and since $\mathbf{A}$ is orthogonal, the inverse transformation is $\mathbf{x} = \mathbf{Az}$. Substituting into (2.2.1) gives

$$(\mathbf{Az})'\mathbf{\Sigma}^{-1}(\mathbf{Az}) = \text{const.}$$

$$= \mathbf{z'A'\Sigma}^{-1}\mathbf{Az}.$$

It is well known that the eigenvectors of $\mathbf{\Sigma}^{-1}$ are the same as those of $\mathbf{\Sigma}$, and that the eigenvalues of $\mathbf{\Sigma}^{-1}$ are the reciprocals of those of $\mathbf{\Sigma}$, assuming that they are all strictly positive. It therefore follows, from a corresponding result to (2.1.3), that $\mathbf{A'\Sigma^{-1}A = \Lambda^{-1}}$ and hence

$$\mathbf{z'\Lambda^{-1}z} = \text{const.}$$

This last equation can be rewritten

$$\sum_{k=1}^{p} \frac{z_k^2}{\lambda_k} = \text{const.} \qquad (2.2.2)$$

and (2.2.2) is the equation for an ellipsoid referred to its principal axes. Equation (2.2.2) also implies that the half-lengths of the principal axes are proportional to $\lambda_1^{1/2}, \lambda_2^{1/2}, \ldots, \lambda_p^{1/2}$.

This result is statistically important if the random vector $\mathbf{x}$ has a multivariate normal distribution, as it quite often assumed. In this case, the ellipsoids given by (2.2.1) define contours of constant probability for the distribution of $\mathbf{x}$. The first (largest) principal axis of such ellipsoids will then define the direction in which statistical variation is greatest, which is another way of expressing the algebraic definition of the first PC given in Section 1.1. The direction of the first PC, defining the first principal axis of constant-probability ellipsoids, is illustrated in Figures 2.1 and 2.2 in Section 2.3. The second principal axis maximizes statistical variation, subject to being orthogonal to the first, and so on, again corresponding to the algebraic definition. This interpretation of PCs, as defining the principal axes of ellipsoids of constant density, was mentioned by Hotelling (1933) in his original paper.

It would appear that this particular geometric property is only of direct statistical relevance if the distribution of $\mathbf{x}$ is multivariate normal, whereas for most other properties of PCs no distributional assumptions are required. However, the property will be discussed further in connection with Property G5 in Section 3.2, where we shall see that it has some relevance even without the assumption of multivariate normality. ☐

Property G2. *Suppose that* $\mathbf{x}_1, \mathbf{x}_2$ *are independent random vectors, both having the same probability distribution, and that* $\mathbf{x}_1, \mathbf{x}_2$ *are both subjected to the same linear transformation*

$$\mathbf{y}_i = \mathbf{B'x}_i, \qquad i = 1, 2.$$

If $\mathbf{B}$ *is a* $(p \times q)$ *matrix with orthonormal columns, chosen to maximize* $E[(\mathbf{y}_1 - \mathbf{y}_2)'(\mathbf{y}_1 - \mathbf{y}_2)]$, *then* $\mathbf{B} = \mathbf{A}_q$, *using the same notation as before.*

PROOF. This result could be viewed as a purely algebraic property, and indeed, the proof below is algebraic. The property is, however, included in the present section because it has a geometric interpretation. This is that the expected squared Euclidean distance, in a q-dimensional subspace, between two vectors of p random variables with the same distribution, is made as large as possible if the subspace is defined by the first q PCs. ☐

To prove Property G2, first note that $\mathbf{x}_1, \mathbf{x}_2$ have the same mean $\mathbf{\mu}$, and covariance matrix $\mathbf{\Sigma}$. Hence $\mathbf{y}_1, \mathbf{y}_2$ also have the same mean and covariance matrix, $\mathbf{B'\mu}, \mathbf{B'\Sigma B}$ respectively.

$$E[(\mathbf{y}_1 - \mathbf{y}_2)'(\mathbf{y}_1 - \mathbf{y}_2)] = E\{[(\mathbf{y}_1 - \mathbf{B}'\boldsymbol{\mu}) - (\mathbf{y}_2 - \mathbf{B}'\boldsymbol{\mu})]'[(\mathbf{y}_1 - \mathbf{B}'\boldsymbol{\mu})$$
$$- (\mathbf{y}_2 - \mathbf{B}'\boldsymbol{\mu})]\}$$
$$= E[(\mathbf{y}_1 - \mathbf{B}'\boldsymbol{\mu})'(\mathbf{y}_1 - \mathbf{B}'\boldsymbol{\mu})]$$
$$+ E[(\mathbf{y}_2 - \mathbf{B}'\boldsymbol{\mu})'(\mathbf{y}_2 - \mathbf{B}'\boldsymbol{\mu})]$$

(the cross-product terms disappear because of the independence of $\mathbf{x}_1$, $\mathbf{x}_2$ and hence of $\mathbf{y}_1$, $\mathbf{y}_2$).

Now, for $i = 1, 2$, we have

$$E[(\mathbf{y}_i - \mathbf{B}'\boldsymbol{\mu})'(\mathbf{y}_i - \mathbf{B}'\boldsymbol{\mu})] = E\{\text{tr}[(\mathbf{y}_i - \mathbf{B}'\boldsymbol{\mu})'(\mathbf{y}_i - \mathbf{B}'\boldsymbol{\mu})]\}$$
$$= E\{\text{tr}[(\mathbf{y}_i - \mathbf{B}'\boldsymbol{\mu})(\mathbf{y}_i - \mathbf{B}'\boldsymbol{\mu})']\}$$
$$= \text{tr}\{E[(\mathbf{y}_i - \mathbf{B}'\boldsymbol{\mu})(\mathbf{y}_i - \mathbf{B}'\boldsymbol{\mu})']\}$$
$$= \text{tr}(\mathbf{B}'\boldsymbol{\Sigma}\mathbf{B}).$$

But $\text{tr}(\mathbf{B}'\boldsymbol{\Sigma}\mathbf{B})$ is maximized when $\mathbf{B} = \mathbf{A}_q$, from Property A1, and the present criterion has been shown above to be $2\,\text{tr}(\mathbf{B}'\boldsymbol{\Sigma}\mathbf{B})$. Hence Property G2 is proved.   □

There is a closely related property, whose geometric interpretation is more tenuous, namely that, with the same definitions as in Property G2,

$$\det\{E[(\mathbf{y}_1 - \mathbf{y}_2)(\mathbf{y}_1 - \mathbf{y}_2)']\}$$

is maximized when $\mathbf{B} = \mathbf{A}_q$—see McCabe (1984). This property says that $\mathbf{B} = \mathbf{A}_q$ makes the generalized variance of $\mathbf{y}_1 - \mathbf{y}_2$ as large as possible, and generalized variance may be viewed as an alternative measure of distance apart of $\mathbf{y}_1$ and $\mathbf{y}_2$ in q-dimensional space (though a less intuitively obvious measure than expected squared Euclidean distance).

Finally, Property G2 can be reversed in the sense that if $E[(\mathbf{y}_1 - \mathbf{y}_2)'(\mathbf{y}_1 - \mathbf{y}_2)]$ or $\det\{E[(\mathbf{y}_1 - \mathbf{y}_2)(\mathbf{y}_1 - \mathbf{y}_2)']\}$ is to be minimized, then this is achieved by taking $\mathbf{B} = \mathbf{A}_q^*$.

The properties given in this section, and in the previous one, show that PCs satisfy several different optimality criteria, but the list of criteria covered is by no means exhaustive—see, for example, Hudlet and Johnson (1982), McCabe (1984) and Okamoto (1969).

2.3. Principal Components Using a Correlation Matrix

The derivation and properties of PCs considered above have been based on the eigenvectors and eigenvalues of the covariance matrix. In practice, as will be seen in much of the remainder of this text, it is more usual to define principal components as

$$z = A'x^*, \qquad (2.3.1)$$

where A now has columns consisting of the eigenvectors of the correlation matrix, and x^* consists of standardized variables. What is being done in adopting such an approach is to find the principal components of a standardized version x^* of x, where x^* has jth element $x_j/\sigma_{jj}^{1/2}, j = 1, 2, \ldots, p, x_j$ is the jth element of x, and σ_{jj} is the variance of x_j. Then the covariance matrix for x^* is the correlation matrix of x, and the PCs of x^* are given by (2.3.1).

A third possibility, instead of using covariance or correlation matrices, is to use covariances of x_j/w_j, where the weights, w_j, are chosen to reflect some *a priori* idea of the relative importance of the variables. The special case $w_j = \sigma_{jj}^{1/2}$ leads to x^*, and to PCs based on the correlation matrix, but various authors have argued that the choice of $w_j = \sigma_{jj}^{1/2}$ is somewhat arbitrary, and that different values of w_j might be better in some applications—see Section 12.1. In practice, however, it must be rare that a uniquely appropriate set of w_j's suggests itself.

All the properties of the previous two sections are still valid for correlation matrices (or indeed for covariances based on other sets of weights), except that we are now considering PCs of x^* (or some other transformation of x), instead of x.

It might seem that the PCs for a correlation matrix could be obtained fairly easily from those for the corresponding covariance matrix, since x^* is related to x by a very simple transformation. However, this is not the case; the eigenvalues and eigenvectors of the correlation matrix have no simple relationship with those of the corresponding covariance matrix. In particular, if the PCs found from the correlation matrix are expressed in terms of x, by transforming back from x^* to x, then these PCs are not the same as the PCs found from Σ, except in very special circumstances (Chatfield and Collins, 1980, Section 4.4). The PCs for correlation and covariance matrices do not, therefore, give equivalent information, nor can they be derived directly from each other, and we now discuss the relative merits of the two types of PC.

A major argument for using correlation matrices, rather than covariance matrices, to define PCs is that the results of analyses for different sets of random variables are more directly comparable than for analyses based on covariance matrices. The big drawback of PCA based on covariance matrices is the sensitivity of the PCs to the units of measurement used for each element of x. If there are large differences between the variances of the elements of x, then those variables whose variances are largest will tend to dominate the first few PCs—see, for example, Section 3.3. This may be entirely appropriate if all the elements of x are measured in the same units, for example, if all elements of x are anatomical measurements on a particular species of animal, all recorded in centimetres, say. Even in such examples, arguments can be presented for the use of correlation matrices (see Section 4.1). In practice, it often occurs that different elements of x are completely different types of measurement. Some might be lengths, some weights, some temperatures, some arbitrary scores on a five-point scale, and so on. In such a case, the structure of the PCs will depend on the choice of units of measurement, as is illustrated by the following artificial example.

Suppose that we have just two variables, x_1, x_2, and that x_1 can equally well be measured in centimetres or in millimetres. The covariance matrices in the two cases are, respectively,

$$\Sigma_1 = \begin{pmatrix} 80 & 44 \\ 44 & 80 \end{pmatrix} \quad \text{and} \quad \Sigma_2 = \begin{pmatrix} 8000 & 440 \\ 440 & 80 \end{pmatrix}.$$

The first PC is $0.707x_1 + 0.707x_2$ for Σ_1 and $0.998x_1 + 0.055x_2$ for Σ_2, so that a relatively minor change in one variable has had the effect of changing a PC which gives equal weight to x_1 and x_2 to a PC which is almost entirely dominated by x_1. Furthermore, the first PC accounts for 77.5% of the total variation for Σ_1, but 99.3% for Σ_2.

Figures 2.1 and 2.2 provide another way of looking at the differences between PCs for the two scales of measurement in x_1. The plots give contours of constant probability (assuming multivariate normality for x) for Σ_1

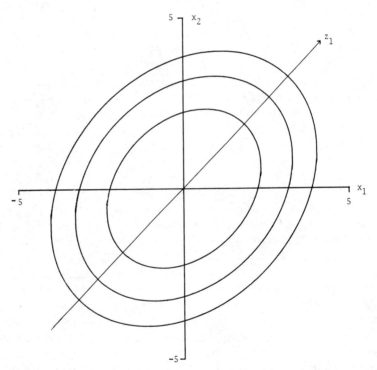

Figure 2.1. Contours of constant probability based on $\Sigma_1 = \begin{pmatrix} 80 & 44 \\ 44 & 80 \end{pmatrix}$.

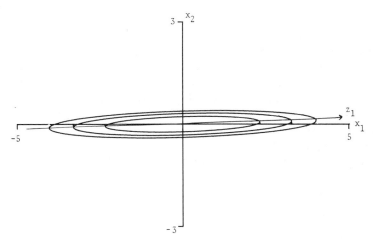

Figure 2.2. Contours of constant probability based on $\Sigma_2 = \begin{pmatrix} 8000 & 440 \\ 440 & 80 \end{pmatrix}$.

and Σ_2 respectively. It is clear from these figures that, whereas with Σ_1 both variables have the same degree of variation, for Σ_2 most of the variation is in the direction of x_1. This is reflected in the first PC which, from Property G1, is defined by the major axis of the ellipses of constant probability.

This example demonstrates the general behaviour of PCs for a covariance matrix when the variances of the individual variables are widely different; the same type of behaviour is illustrated again, for samples, in Section 3.3. The first PC is dominated by the variable with the largest variance, the second PC is dominated by the variable with the second largest variance, and so on, with a substantial proportion of the total variation accounted for by just two or three PCs. In other words, the PCs are little different from the original variables, rearranged in decreasing order of the size of their variances. Also, the first few PCs account for little of the off-diagonal elements of Σ in this case—see Property A3 above. In most circumstances, such a transformation is of little value, and it will not occur if the correlation, rather than covariance, matrix is used.

The example has shown that it is unwise to use PCs on a covariance matrix when x consists of measurements of different types, unless there is a strong conviction that the units of measurements chosen for each element of x are the only ones which make sense. Even if this condition holds, using the covariance matrix will not provide very informative PCs if the variables have widely differing variances.

Another problem with the use of covariance matrices is that it is more difficult to compare informally the results from different analyses than with correlation matrices. Sizes of variances of PCs have the same implications for different correlation matrices, but not for different covariance matrices. Also, patterns of coefficients in PCs can be readily compared for different correlation matrices to see if the two correlation matrices are giving similar PCs, whereas informal comparisons are often much trickier for covariance matrices. Formal methods for comparing PCs from different covariance matrices are, however, available—see Section 11.5.

The use of covariance matrices does have one general advantage over correlation matrices, and one advantage which occurs in a special case. The general advantage is that statistical inference regarding population PCs, based on sample PCs, is easier for covariance matrices, as will be discussed in Section 3.7. This is an important advantage if statistical inference is thought to play a crucial role in PCA. However, in practice, it is more common to use PCA as a descriptive, rather than inferential, tool, so that the advantage becomes less crucial.

The second advantage of covariance matrices holds in the special case when all elements of x are measured in the same units. It can then be argued that standardizing the elements of x to give correlations is equivalent to making an arbitrary choice of measurement units. This argument of arbitrariness can also be applied more generally to the use of correlation matrices, but when the elements of x are measurements of different types, the choice of measurement units leading to a covariance matrix is even more arbitrary, so that the correlation matrix is preferred.

We conclude this section by looking at two interesting properties which hold for PCs derived from the correlation matrix. The first is that the PCs depend not on the absolute values of correlations, but only on their ratios. This follows because multiplication of all off-diagonal elements of a correlation matrix by the same constant leaves the eigenvectors of the matrix unchanged (Chatfield and Collins, 1980, p. 67).

The second property, which was noted by Hotelling (1933) in his original paper, is that if, instead of the normalization $\alpha'_k \alpha_k = 1$, we use

$$\tilde{\alpha}'_k \tilde{\alpha}_k = \lambda_k, \qquad k = 1, 2, \ldots, p, \tag{2.3.2}$$

then $\tilde{\alpha}_{kj}$, the jth element of $\tilde{\alpha}_k$, is the correlation between the jth variable and the kth PC. To see this note that, for $k = 1, 2, \ldots, p$,

$$\tilde{\alpha}_k = \lambda_k^{1/2} \alpha_k, \qquad \mathrm{var}(z_k) = \lambda_k,$$

and the p-element vector $\Sigma \alpha_k$ has as its jth element the covariance between x_j and z_k. But $\Sigma \alpha_k = \lambda_k \alpha_k$, so the covariance between x_j and z_k is $\lambda_k \alpha_{kj}$. The correlation between x_j and z_k is therefore

$$\frac{\lambda_k \alpha_{kj}}{[\mathrm{var}(x_j)\,\mathrm{var}(z_k)]^{1/2}} = \frac{\lambda_k^{1/2} \alpha_{kj}}{\sigma_{jj}^{1/2}}$$

$$= \tilde{\alpha}_{kj}/\sigma_{jj}^{1/2},$$

where $\sigma_{jj} = \mathrm{var}(x_j)$. In the case where PCs are found from a correlation matrix, $\sigma_{jj} = \mathrm{var}(x_j^*) = 1$, so that $\tilde{\alpha}_{kj}$ is indeed the correlation between the kth PC and the jth variable.

Because of this property the normalization (2.3.2) is quite often used, in particular in computer packages (see Appendix A2), but it has the disadvantage that it is less easy to informally interpret and compare a set of PCs when each PC has a different normalization on its coefficients.

Both of the above properties hold for correlation matrices, and, although similar ideas can be examined for covariance matrices, the results are, in each case, less straightforward.

2.4. Principal Components with Equal and/or Zero Variances

The final, short, section of this chapter discusses two problems which can arise in theory, but are relatively uncommon in practice. In most of this chapter it has been assumed, implicitly or explicitly, that the eigenvalues of the covariance (or correlation) matrix are all different, and that none of them are zero.

Equality of eigenvalues, and hence equality of variances of PCs will occur for certain patterned matrices. The effect of this occurrence is that for a group of q equal eigenvalues, the corresponding q eigenvectors span a certain unique q-dimensional space, but, within this space, they are, apart from being orthogonal to one another, arbitrary. Geometrically, (see Property G1), what is happening for $q = 2$ or 3, is that the principal axes of a circle or sphere cannot be uniquely defined; a similar problem arises for hyperspheres when $q > 3$. Thus individual PCs corresponding to eigenvalues in a group of equal eigenvalues, are not uniquely defined. A further problem with equal-variance PCs is that statistical inference becomes more complicated (see Section 3.7). In practice, we need not worry too much about equal eigenvalues. It will be relatively rare for a population covariance structure to have exactly equal eigenvalues, and even more unusual for equality to hold for sample covariances.

The other complication, variances equal to zero, occurs rather more frequently, but is still fairly unusual. If q eigenvalues are zero, then the rank of Σ is $p - q$, rather than p, and this outcome necessitates modifications to the proofs of some properties given in Section 2.1 above. Any PC with zero variance defines an exactly constant linear relationship between the elements of $\mathbf{x}$. If such relationships exist, then they imply that one variable is redundant for each relationship, since its value can be determined exactly from the values of the other variables which appear in the relationship. We could therefore reduce the number of variables from p to $p - q$ without losing any information. Ideally, exact linear relationships should be spotted before doing a PCA, and the number of variables reduced accordingly. Alternatively, any exact, or near-exact, linear relationships uncovered by the last few PCs can be used to select a subset of variables which contain most of the information available in all of the original variables. This idea is discussed further in Section 6.3.

There will always be the same number of zero eigenvalues for a correlation matrix as for the corresponding covariance matrix, since an exact linear relationship between the elements of x clearly implies an exact linear relationship between the standardized variables, and vice versa. There is not the same equivalence, however, when it comes to considering equal variance PCs. Equality of some of the eigenvalues in a covariance (correlation) matrix need not imply that any of the eigenvalues of the corresponding correlation (covariance) matrix are equal. A simple example is when the p variables all have equal correlations but unequal variances. If $p > 2$, then the last $p - 1$ eigenvalues of the correlation matrix are equal (see Morrison, 1976, Section 8.6), but this relationship will not hold, in general, for the covariance matrix. Further discussion of patterns in covariance or correlation matrices, and their implications for the structure of the corresponding PCs, is given in Section 3.8.

CHAPTER 3

Mathematical and Statistical Properties of Sample Principal Components

The first part of this chapter will be similar in structure to Chapter 2, except that it will deal with properties of PCs obtained from a sample covariance (or correlation) matrix, rather than from a population covariance (or correlation) matrix. The first two sections of the chapter, as in Chapter 2, describe respectively many of the algebraic and geometric properties of PCs. Most of the properties discussed in Chapter 2 are almost the same for samples as for populations, and will only be mentioned again briefly. There are, however, some additional properties which are relevant only to sample PCs and these will be discussed more fully.

The third and fourth sections of the chapter again mirror those of Chapter 2. The third section discusses again, with an example, the choice between correlation and covariance matrices, while the fourth section looks at the implications of equal and/or zero variances among the PCs, and illustrates the potential usefulness of the last few PCs in detecting near-constant relationships between the variables.

The last four sections of the chapter cover material which has no counterpart in Chapter 2. Section 3.5 discusses the singular value decomposition, which could have been included in Section 3.1 as an additional algebraic property. However, it is sufficiently important to warrant its own section, since it provides a useful alternative approach to some of the theory surrounding PCs, as well as giving an efficient practical method for actually computing PCs.

The sixth section looks at the probability distributions of the coefficients and the variances of a set of sample PCs, in other words, the probability distributions of the eigenvectors and eigenvalues of a sample covariance matrix. The seventh section then goes on to show how these distributions may be used to make statistical inferences about the population PCs, based on sample PCs.

Finally, Section 3.8 discusses how the approximate structure, and variances, of PCs can sometimes be deduced from patterns in the covariance or correlation matrix.

3.1. Optimal Algebraic Properties of Sample Principal Components

Before looking at the properties themselves, we need to establish some notation. Suppose that we have n independent observations on the p-element random vector x; denote these n observations by $x_1, x_2, \ldots, x_n$. Let $\tilde{z}_{i1} = a_1' x_i$, $i = 1, 2, \ldots, n$, and choose the vector of coefficients a_1' to maximize the sample variance $[1/(n-1)] \sum_{i=1}^{n} (\tilde{z}_{i1} - \bar{z}_1)^2$, subject to the normalization constraint $a_1' a_1 = 1$. Next let $\tilde{z}_{i2} = a_2' x_i$, $i = 1, 2, \ldots, n$, and choose a_2 to maximize the sample variance of the $\tilde{z}_{i2}$'s, subject to the normalization constraint $a_2' a_2 = 1$, and subject also to the $\tilde{z}_{i2}$'s being uncorrelated with $\tilde{z}_{i1}$'s in the sample. Continuing this process in an obvious manner, we have a sample version of the definition of PCs which was given in Section 1.1. $a_k' x$ is therefore defined as the kth sample PC, $k = 1, 2, \ldots, p$, and $\tilde{z}_{ik}$ is the *score* for the ith observation on the kth PC. If the derivation in Section 1.1 is followed through, but with sample variances and covariances replacing population quantities, then it turns out that the sample variance of the PC scores for the kth sample PC is l_k, the kth largest eigenvalue of S (the sample covariance matrix for $x_1, x_2, \ldots, x_n$) and a_k is the corresponding eigenvector, for $k = 1, 2, \ldots, p$.

Define the $(n \times p)$ matrices $\tilde{X}$ and $\tilde{Z}$ to have (i, k)th elements equal to the values of the kth element of x_i, and to $\tilde{z}_{ik}$ respectively. Then $\tilde{Z}$ and $\tilde{X}$ are related by $\tilde{Z} = \tilde{X}A$ where A is the $(p \times p)$ orthogonal matrix, whose kth column is a_k.

If the mean of each element of x is known to be zero, then $S = (1/n)\tilde{X}'\tilde{X}$. It is far more usual for the mean of x to be unknown, and in this case the (j, k)th element of S is

$$\frac{1}{(n-1)} \sum_{i=1}^{n} (x_{ij} - \bar{x}_j)(x_{ik} - \bar{x}_k),$$

where

$$\bar{x}_j = \frac{1}{n} \sum_{i=1}^{n} x_{ij}, \qquad j, k = 1, 2, \ldots, p.$$

The matrix S can therefore be written as

$$S = \frac{1}{(n-1)} X'X, \qquad (3.1.1)$$

where X is an $(n \times p)$ matrix with (i, j)th element $(x_{ij} - \bar{x}_j)$; the representation

(3.1.1) will be very useful in this and subsequent chapters. In addition, it will be convenient in some places below to use the notation x_{ij} to denote the (i, j)th element of $\mathbf{X}$, so that x_{ij} will be the value of the jth variable, *measured about its mean,* $\bar{x}_j$, for the ith observation. A final notational point is that it will be convenient to define the matrix of PC scores as

$$\mathbf{Z} = \mathbf{XA}, \qquad (3.1.2)$$

rather than the earlier definition. These PC scores will have exactly the same variances and covariances as those given by $\tilde{\mathbf{Z}}$, but will have zero means, rather than means $\bar{z}_k, k = 1, 2, \ldots, p$.

Another point to note is that the eigenvectors of $[1/(n-1)]\mathbf{X'X}$ and $\mathbf{X'X}$ are identical, and the eigenvalues of $[1/(n-1)]\mathbf{X'X}$ are simply $[1/(n-1)]$ (the eigenvalues of $\mathbf{X'X}$). Because of these relationships it will be convenient in some places below to work in terms of eigenvalues and eigenvectors of $\mathbf{X'X}$, rather than directly with those of $\mathbf{S}$.

Turning to the algebraic properties A1–A5 listed in Section 2.1, define

$$\mathbf{y}_i = \mathbf{B'x}_i \qquad \text{for } i = 1, 2, \ldots, n, \qquad (3.1.3)$$

where $\mathbf{B}$, as in Properties A1, A2, A4, A5, is a $(p \times q)$ matrix whose columns are orthonormal. Then Properties A1, A2, A4, A5, still hold, but with the sample covariance matrix of the observations $\mathbf{y}_i, i = 1, 2, \ldots, n$ replacing Σ_y, and with the matrix $\mathbf{A}$ now defined as having kth column $\mathbf{a}_k$, with $\mathbf{A}_q, \mathbf{A}_q^*$ respectively representing its first and last q columns. Proofs in all cases are similar to those for populations, after making appropriate substitutions of sample, in place of population, quantities, and will not be repeated. Property A5 reappears as Property G3 in the next section and a proof will be given there.

The spectral decomposition, Property A3, also holds for samples, in the form

$$\mathbf{S} = l_1\mathbf{a}_1\mathbf{a}_1' + l_2\mathbf{a}_2\mathbf{a}_2' + \cdots + l_p\mathbf{a}_p\mathbf{a}_p'. \qquad (3.1.4)$$

The statistical implications of this expression, and the other algebraic properties, A1, A2, A4, A5, are virtually the same as for the corresponding population properties in Section 2.1, except that they must now be viewed in a sample context.

One further reason can be put forward for interest in the last few PCs, as found by Property A2, in the case of sample correlation matrices. Raveh (1985) argues that the inverse, $\mathbf{R}^{-1}$, of a correlation matrix is of greater interest, in many situations, than $\mathbf{R}$. It may therefore be more important to approximate $\mathbf{R}^{-1}$, then $\mathbf{R}$, in a few dimensions. If this is done using the spectral decomposition (Property A3), of $\mathbf{R}^{-1}$, then the first few terms will correspond to the last few PCs, since eigenvectors of $\mathbf{R}, \mathbf{R}^{-1}$ are the same, except that their order is reversed. The rôle of the last few PCs will be discussed further in Sections 3.4 and 3.7, and again in Sections 6.3, 8.4, 8.6 and 10.1.

We shall deal with one other property, which is concerned with the use of principal components in regression, in this section. Standard terminology from regression will be used and will not be explained in detail—see, for example, Draper and Smith (1981). A full discussion of the use of principal components in regression is given in Chapter 8.

Property A6. *Suppose now that* $\mathbf{X}$, *defined as above, consists of n observations on p predictor variables,* $\mathbf{x}$, *measured about their sample means, and that the corresponding regression equation is*

$$\mathbf{y} = \mathbf{X}\boldsymbol{\beta} + \boldsymbol{\varepsilon}, \tag{3.1.5}$$

where $\mathbf{y}$ *is the vector of n observations on the dependent variable, again measured about the sample mean. (The notation* $\mathbf{y}$ *for the dependent variable has no connection with the usage of* $\mathbf{y}$ *elsewhere in the chapter, but is standard in regression.) Suppose that* $\mathbf{X}$ *is transformed by the equation* $\mathbf{Z} = \mathbf{X}\mathbf{B}$, *where* $\mathbf{B}$ *is a* $(p \times p)$ *orthogonal matrix. The regression equation can then be rewritten as*

$$\mathbf{y} = \mathbf{Z}\boldsymbol{\gamma} + \boldsymbol{\varepsilon},$$

where $\boldsymbol{\gamma} = \mathbf{B}^{-1}\boldsymbol{\beta}$. *The usual, least squares, estimator for* $\boldsymbol{\gamma}$ *is* $\hat{\boldsymbol{\gamma}} = (\mathbf{Z}'\mathbf{Z})^{-1}\mathbf{Z}'\mathbf{y}$. *Then the elements of* $\hat{\boldsymbol{\gamma}}$ *have, successively, the smallest possible variances if* $\mathbf{B} = \mathbf{A}$, *the matrix whose* kth *column is the* kth *eigenvector of* $\mathbf{X}'\mathbf{X}$ *(and hence the* kth *eigenvector of* $\mathbf{S}$). $\mathbf{Z}$ *therefore consists of values of the sample principal components for* $\mathbf{x}$.

PROOF. From standard results in regression (Draper and Smith, 1981, Section 2.6) the covariance matrix of the least squares estimator, $\hat{\boldsymbol{\gamma}}$, is proportional to

$$(\mathbf{Z}'\mathbf{Z})^{-1} = (\mathbf{B}'\mathbf{X}'\mathbf{X}\mathbf{B})^{-1} = \mathbf{B}^{-1}(\mathbf{X}'\mathbf{X})^{-1}(\mathbf{B}')^{-1} = \mathbf{B}'(\mathbf{X}'\mathbf{X})^{-1}\mathbf{B},$$

since $\mathbf{B}$ is orthogonal. We require to minimize, for $q = 1, 2, \ldots, p$, $\text{tr}(\mathbf{B}_q'(\mathbf{X}'\mathbf{X})^{-1}\mathbf{B}_q)$, where $\mathbf{B}_q$ consists of the first q columns of $\mathbf{B}$. But, replacing $\boldsymbol{\Sigma}_y$ by $(\mathbf{X}'\mathbf{X})^{-1}$ in Property A2 of Section 2.1 shows that $\mathbf{B}_q$ must consist of the last q columns of a matrix whose kth column is the kth eigenvector of $(\mathbf{X}'\mathbf{X})^{-1}$. Furthermore, $(\mathbf{X}'\mathbf{X})^{-1}$ has the same eigenvectors as $\mathbf{X}'\mathbf{X}$, except that their order is reversed, so that $\mathbf{B}_q$ must have columns equal to the first q eigenvectors of $\mathbf{X}'\mathbf{X}$. Since this holds for $q = 1, 2, \ldots, p$, Property A6 is proved. □

This property seems to imply that replacing the predictor variables in a regression analysis by their first few PCs is an attractive idea, since those PCs omitted have coefficients which are estimated with little precision. The flaw in this argument is that nothing in Property A6 takes account of the strength of the relationship between the dependent variable y and the elements of $\mathbf{x}$, or between y and the PCs. A large variance for $\hat{\gamma}_k$, the kth element of $\boldsymbol{\gamma}$, and hence an imprecise estimate of the degree of relationship between y and the kth PC, z_k, does *not* preclude a strong relationship between y and z_k (see Section 8.2). Further discussion of Property A6 is given by Fomby *et al.* (1978).

There are a number of other properties of PCs specific to the sample situation; most have geometric interpretations and are therefore dealt with in the next section.

3.2. Geometric Properties of Sample Principal Components

As with the algebraic properties, the geometric properties of Chapter 2 are also relevant for sample PCs, although with slight modifications to the statistical implications. In addition to these properties, the present section includes a proof of a sample version of Property A5, viewed geometrically, and introduces two extra properties which are relevant to sample, but not population, PCs.

Property G1 is still valid for samples, if Σ is replaced by S. The ellipsoids $x'S^{-1}x = $ const. no longer have the interpretation of being contours of constant probability, though they will provide estimates of such contours if $x_1, x_2, \ldots, x_n$ are drawn from a multivariate normal distribution.

Property G2 may also be carried over from populations to samples as follows. Suppose that the observations $x_1, x_2, \ldots, x_n$, are transformed by

$$y_i = B'x_i, \qquad i = 1, 2, \ldots, n,$$

where B is a $(p \times q)$ matrix with orthonormal columns, so that $y_1, y_2, \ldots, y_n$ are projections of $x_1, x_2, \ldots, x_n$ onto a q-dimensional subspace. Then $\sum_{h=1}^{n} \sum_{i=1}^{n} (y_h - y_i)'(y_h - y_i)$ is maximized when $B = A_q$. Conversely, the same criterion is minimized when $B = A_q^*$.

This property means that if the n observations are projected onto a q-dimensional subspace, then the sum of squared Euclidean distances between all pairs of observations in the subspace is maximized when the subspace is defined by the first q PCs, and minimized when it is defined by the last q PCs. The proof that this property holds is again rather similar to that for the corresponding population property and will not be repeated.

The next property to be considered is equivalent to Property A5.

Property G3. *As before, suppose that the observations $x_1, x_2, \ldots, x_n$ are transformed by $y_i = B'x_i$, $i = 1, 2, \ldots, n$, where B is a $(p \times q)$ matrix with orthonormal columns, so that $y_1, y_2, \ldots, y_n$ are projections of $x_1, x_2, \ldots, x_n$ onto a q-dimensional subspace. A measure of 'goodness-of-fit' of this q-dimensional subspace to $x_1, x_2, \ldots, x_n$ can be defined as the sum of squared perpendicular distances of $x_1, x_2, \ldots, x_n$ from the subspace. This measure is minimized when $B = A_q$.*

Proof. y_i is an orthogonal projection of x_i onto a q-dimensional subspace defined by the matrix B. Let m_i denote the position of y_i in terms of the

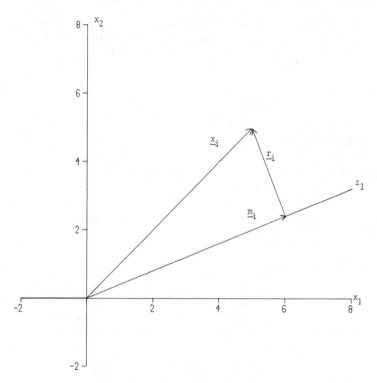

Figure 3.1. Orthogonal projection of a two-dimensional vector onto a one-dimensional subspace.

original co-ordinates, and $\mathbf{r}_i = \mathbf{x}_i - \mathbf{m}_i$. (See Figure 3.1 for the special case where $p = 2$, $q = 1$—in this case $\mathbf{y}_i$ is a scalar, whose value is the length of $\mathbf{m}_i$.) Because $\mathbf{m}_i$ is an orthogonal projection of $\mathbf{x}_i$ onto a q-dimensional subspace, $\mathbf{r}_i$ is orthogonal to the subspace, so $\mathbf{r}_i'\mathbf{m}_i = 0$. Furthermore, $\mathbf{r}_i'\mathbf{r}_i$ is the squared perpendicular distance of $\mathbf{x}_i$ from the subspace, so that the sum of squared perpendicular distances of $\mathbf{x}_1, \mathbf{x}_2, \ldots, \mathbf{x}_n$ from the subspace is

$$\sum_{i=1}^{n} \mathbf{r}_i'\mathbf{r}_i.$$

Now

$$\mathbf{x}_i'\mathbf{x}_i = (\mathbf{m}_i + \mathbf{r}_i)'(\mathbf{m}_i + \mathbf{r}_i)$$

$$= \mathbf{m}_i'\mathbf{m}_i + \mathbf{r}_i'\mathbf{r}_i + 2\mathbf{r}_i'\mathbf{m}_i$$

$$= \mathbf{m}_i'\mathbf{m}_i + \mathbf{r}_i'\mathbf{r}_i.$$

Thus

$$\sum_{i=1}^{n} \mathbf{r}_i'\mathbf{r}_i = \sum_{i=1}^{n} \mathbf{x}_i'\mathbf{x}_i - \sum_{i=1}^{n} \mathbf{m}_i'\mathbf{m}_i,$$

so that, for a given set of observations, minimization of the sum of squared perpendicular distances is equivalent to maximization of $\sum_{i=1}^{n} \mathbf{m}_i'\mathbf{m}_i$. Distances are preserved under orthogonal transformations, so the squared distance $\mathbf{m}_i'\mathbf{m}_i$ of $\mathbf{y}_i$ from the origin is the same in y co-ordinates as in x co-ordinates. Therefore, the quantity to be maximized is $\sum_{i=1}^{n} \mathbf{y}_i'\mathbf{y}_i$. But

$$\sum_{i=1}^{n} \mathbf{y}_i'\mathbf{y}_i = \sum_{i=1}^{n} \mathbf{x}_i'\mathbf{BB}'\mathbf{x}_i$$

$$= \text{tr} \sum_{i=1}^{n} (\mathbf{x}_i'\mathbf{BB}'\mathbf{x}_i)$$

$$= \sum_{i=1}^{n} \text{tr}(\mathbf{x}_i'\mathbf{BB}'\mathbf{x}_i)$$

$$= \sum_{i=1}^{n} \text{tr}(\mathbf{B}'\mathbf{x}_i\mathbf{x}_i'\mathbf{B})$$

$$= \text{tr}\left[\mathbf{B}' \left(\sum_{i=1}^{n} \mathbf{x}_i\mathbf{x}_i' \right) \mathbf{B} \right]$$

$$= \text{tr}[\mathbf{B}'\mathbf{X}'\mathbf{XB}]$$

$$= (n - 1)\, \text{tr}(\mathbf{B}'\mathbf{SB}).$$

But from Property A1, $\text{tr}(\mathbf{B}'\mathbf{SB})$ is maximized when $\mathbf{B} = \mathbf{A}_q$.

Instead of treating this property (G3) as just another property of sample PCs, it can also be viewed as an alternative derivation of the PCs. Rather than adapting for samples the algebraic definition of population PCs given in Chapter 1, there is an alternative geometric definition of sample PCs. They are defined as the linear functions (projections) of $\mathbf{x}_1, \mathbf{x}_2, \ldots, \mathbf{x}_n$, which successively define subspaces of dimension $1, 2, \ldots, q, \ldots, p - 1$, for which the sum of squared perpendicular distances of $\mathbf{x}_1, \mathbf{x}_2, \ldots, \mathbf{x}_n$ from the subspace is minimized. In fact, this is essentially the approach adopted by Pearson (1901), although he concentrated on the two special cases where $q = 1$ and $q = p - 1$. Given a set of points in p-dimensional space, Pearson found the 'best-fitting line', and the 'best-fitting ($q - 1$ dimensional hyper-) plane', in the sense of minimizing the sum of squared deviations of the points from the line or plane. The best-fitting line determines the first PC (although Pearson did not use this terminology), and the direction of the last PC is orthogonal to the best-fitting plane. The scores for the last PC are simply the perpendicular distances of the observations from this best-fitting hyperplane. □

Property G4. *Let $\mathbf{X}$ be the $(n \times p)$ matrix whose (i, j)th element is $x_{ij} - \bar{x}_j$, and consider the matrix $\mathbf{XX}'$. The ith diagonal element of $\mathbf{XX}'$ is $\sum_{j=1}^{p}(x_{ij} - \bar{x}_j)^2$, which is the squared Euclidean distance of $\mathbf{x}_i$ from the centre of gravity, $\bar{\mathbf{x}}$, of the points $\mathbf{x}_1, \mathbf{x}_2, \ldots, \mathbf{x}_n$ $(\bar{\mathbf{x}} = (1/n)\sum_{i=1}^{n} \mathbf{x}_i)$. Also, the (h, i)th element of $\mathbf{XX}'$ is*

$\sum_{j=1}^{p} (x_{hj} - \bar{x}_j)(x_{ij} - \bar{x}_j)$, which measures the cosine of the angle between the lines joining $\mathbf{x}_h$ and $\mathbf{x}_i$ to $\bar{\mathbf{x}}$, multiplied by their distances from $\bar{\mathbf{x}}$. Thus $\mathbf{XX}'$ contains information about the configuration of $\mathbf{x}_1, \mathbf{x}_2, \ldots, \mathbf{x}_n$ relative to $\bar{\mathbf{x}}$. Now suppose that $\mathbf{x}_1, \mathbf{x}_2, \ldots, \mathbf{x}_n$ are projected onto a q-dimensional subspace with the usual orthogonal transformation $\mathbf{y}_i = \mathbf{B}'\mathbf{x}_i$, $i = 1, 2, \ldots, n$. Then, the transformation for which $\mathbf{B} = \mathbf{A}_q$ minimizes the distortion in the configuration, as measured by $\|\mathbf{YY}' - \mathbf{XX}'\|$, where $\|\cdot\|$ denotes Euclidean norm, and $\mathbf{Y}$ is a matrix with (i, j)th element $y_{ij} - \bar{y}_j$.

PROOF. $\mathbf{Y} = \mathbf{XB}$, so

$$\mathbf{YY}' = \mathbf{XBB}'\mathbf{X}' \quad \text{and} \quad \|\mathbf{YY}' - \mathbf{XX}'\| = \|\mathbf{XBB}'\mathbf{X}' - \mathbf{XX}'\|.$$

A matrix result, given by Rao (1973, p. 63), states that if $\mathbf{F}$ is a symmetric matrix of rank p with spectral decomposition

$$\mathbf{F} = f_1\boldsymbol{\phi}_1\boldsymbol{\phi}_1' + f_2\boldsymbol{\phi}_2\boldsymbol{\phi}_2' + \cdots + f_p\boldsymbol{\phi}_p\boldsymbol{\phi}_p',$$

and $\mathbf{G}$ is a matrix of rank $q < p$ chosen to minimize $\|\mathbf{F} - \mathbf{G}\|$ then

$$\mathbf{G} = f_1\boldsymbol{\phi}_1\boldsymbol{\phi}_1' + f_2\boldsymbol{\phi}_2\boldsymbol{\phi}_2' + \cdots + f_q\boldsymbol{\phi}_q\boldsymbol{\phi}_q'.$$

Assuming that $\mathbf{X}$ has rank p (so that $\mathbf{x}_i - \bar{\mathbf{x}}$, $i = 1, 2, \ldots, n$, span p-dimensional space, and are not contained in any proper subspace), then so does $\mathbf{XX}'$, and the above result can be used with $\mathbf{F} = \mathbf{XX}'$, and $\mathbf{G} = \mathbf{YY}'$.

Now, if l_k, $\mathbf{a}_k$ denote the kth eigenvalue and eigenvector respectively of $\mathbf{X}'\mathbf{X}$, then the kth eigenvalue and eigenvector of $\mathbf{XX}'$ are l_k and $l_k^{-1/2}\mathbf{Xa}_k$ respectively (the remaining $(n - p)$ eigenvalues of $\mathbf{XX}'$ are zero).

Using the general result above, $\|\mathbf{YY}' - \mathbf{XX}'\|$ is minimized when

$$\mathbf{G} = \mathbf{XBB}'\mathbf{X}' = l_1^{-1}l_1\mathbf{Xa}_1\mathbf{a}_1'\mathbf{X}' + l_2^{-1}l_2\mathbf{Xa}_2\mathbf{a}_2'\mathbf{X}' + \cdots + l_q^{-1}l_q\mathbf{Xa}_q\mathbf{a}_q'\mathbf{X}',$$

or

$$\mathbf{XBB}'\mathbf{X}' = \mathbf{Xa}_1\mathbf{a}_1'\mathbf{X}' + \mathbf{Xa}_2\mathbf{a}_2'\mathbf{X}' + \cdots + \mathbf{Xa}_q\mathbf{a}_q'\mathbf{X}'.$$

Multiplying both sides of this equation on the left by $(\mathbf{X}'\mathbf{X})^{-1}\mathbf{X}'$ and on the right by $\mathbf{X}(\mathbf{X}'\mathbf{X})^{-1}$, gives

$$\mathbf{BB}' = \mathbf{a}_1\mathbf{a}_1' + \mathbf{a}_2\mathbf{a}_2' + \cdots + \mathbf{a}_q\mathbf{a}_q',$$

from which it follows that the columns of $\mathbf{B}$ and the first q eigenvectors of $\mathbf{X}'\mathbf{X}$ (or equivalently of $\mathbf{S}$), span the same q-dimensional subspace. In other words the transformation $\mathbf{B} = \mathbf{A}_q$ provides the required optimal subspace. $\quad\square$

Note that the result given by Rao (1973, p. 63), which was used in the above proof, implies that the sum of the first q terms in the spectral decomposition of the sample covariance (or correlation) matrix, $\mathbf{S}$, provides the rank q matrix ${}_q\mathbf{S}$ which minimizes $\|{}_q\mathbf{S} - \mathbf{S}\|$. Furthermore, $\|{}_q\mathbf{S} - \mathbf{S}\| = \sum_{k=q+1}^{p} l_k$, where l_k now denotes the kth eigenvalue of $\mathbf{S}$, rather than that of $\mathbf{X}'\mathbf{X}$. The result follows because

$$\|_q\mathbf{S} - \mathbf{S}\| = \left\| \sum_{k=q+1}^{p} l_k \mathbf{a}_k \mathbf{a}_k' \right\|$$

$$= \sum_{k=q+1}^{p} l_k \|\mathbf{a}_k \mathbf{a}_k'\|$$

$$= \sum_{k=q+1}^{p} l_k \left[\sum_{i=1}^{p} \sum_{j=1}^{p} (a_{ki} a_{kj})^2 \right]^{1/2}$$

$$= \sum_{k=q+1}^{p} l_k \left[\sum_{i=1}^{p} a_{ki}^2 \sum_{j=1}^{p} a_{kj}^2 \right]^{1/2}$$

$$= \sum_{k=q+1}^{p} l_k,$$

because $\mathbf{a}_k' \mathbf{a}_k = 1$, $k = 1, 2, \ldots, p$.

Property G4 is very similar to another optimality property of PCs, discussed in terms of the so-called RV-coefficient by Robert and Escoufier (1976). The RV-coefficient was introduced as a measure of the similarity between two configurations of n data points, as described by $\mathbf{XX}'$ and $\mathbf{YY}'$. The distance between the two configurations is defined by Robert and Escoufier (1976) as

$$\left\| \frac{\mathbf{XX}'}{\{\mathrm{tr}(\mathbf{XX}')^2\}^{1/2}} - \frac{\mathbf{YY}'}{\{\mathrm{tr}(\mathbf{YY}')^2\}^{1/2}} \right\|, \tag{3.2.1}$$

where the divisors of $\mathbf{XX}'$, $\mathbf{YY}'$ are introduced simply to standardize the representation of each configuration in the sense that

$$\left\| \frac{\mathbf{XX}'}{\{\mathrm{tr}(\mathbf{XX}')^2\}^{1/2}} \right\| = \left\| \frac{\mathbf{YY}'}{\{\mathrm{tr}(\mathbf{YY}')^2\}^{1/2}} \right\| = 1.$$

It can then be shown that (3.2.1) equals $[2(1 - \mathrm{RV}(\mathbf{X}, \mathbf{Y}))]^{1/2}$, where the RV-coefficient is defined as

$$\mathrm{RV}(\mathbf{X}, \mathbf{Y}) = \frac{\mathrm{tr}(\mathbf{XY}'\mathbf{YX}')}{\{\mathrm{tr}(\mathbf{XX}')^2 \, \mathrm{tr}(\mathbf{YY}')^2\}^{1/2}}. \tag{3.2.2}$$

Thus, minimizing the distance measure (3.2.1) which, apart from standardizations, is the same as the criterion of Property G4, is equivalent to maximization of $\mathrm{RV}(\mathbf{X}, \mathbf{Y})$. Robert and Escoufier (1976) show that several multivariate techniques can be expressed in terms of maximizing $\mathrm{RV}(\mathbf{X}, \mathbf{Y})$ for some definition of $\mathbf{X}$ and $\mathbf{Y}$. In particular, if $\mathbf{Y}$ is restricted to be of the form $\mathbf{Y} = \mathbf{XB}$, where $\mathbf{B}$ is a $(p \times q)$ matrix, such that the columns of $\mathbf{Y}$ are uncorrelated, then maximization of $\mathrm{RV}(\mathbf{X}, \mathbf{Y})$ leads to $\mathbf{B} = \mathbf{A}_q$, i.e. $\mathbf{Y}$ consists of scores on the first q PCs.

Property G5. *The algebraic derivation of sample PCs reduces to finding, successively, vectors* $\mathbf{a}_k$, $k = 1, 2, \ldots, p$, *which maximize* $\mathbf{a}_k' \mathbf{S} \mathbf{a}_k$ *subject to* $\mathbf{a}_k' \mathbf{a}_k = 1$, *and subject to* $\mathbf{a}'_k\mathbf{a}_l = 0$ *for* $l < k$. *This statement of the problem can be viewed geometrically as follows* (*Stuart*, 1982).

Consider the first PC; this maximizes $\mathbf{a}'\mathbf{Sa}$ subject to $\mathbf{a}'\mathbf{a} = 1$. But $\mathbf{a}'\mathbf{Sa} = $ const. defines a family of ellipsoids and $\mathbf{a}'\mathbf{a} = 1$ defines a hypersphere in p-dimensional space, both centred at the origin. The hypersphere $\mathbf{a}'\mathbf{a} = 1$ will intersect more than one of the ellipsoids in the family $\mathbf{a}'\mathbf{Sa}$ (*unless* **S** *is the identity matrix*), and the points at which the hypersphere intersects the '*biggest*' such ellipsoid (so that $\mathbf{a}'\mathbf{Sa}$ is maximized) lie on the shortest principal axis of the ellipse. A simple diagram, as given by Stuart (1982), readily verifies this result when $p = 2$. The argument can be extended to show that the first q sample PCs are defined by the q shortest principal axes of the family of ellipsoids $\mathbf{a}'\mathbf{Sa} = $ const. Although Stuart (1982) introduced this interpretation in terms of sample PCs, it is equally valid for population PCs.

The earlier geometric property G1 was also concerned with ellipsoids, but in the context of multivariate normality where the ellipsoids $\mathbf{x}'\mathbf{\Sigma}^{-1}\mathbf{x} = $ const. define contours of constant probability, and where the first (longest) q principal axes of such ellipsoids define the first q population PCs. In the light of Property G5, it is clear that the validity of the result G1 does not really depend on the assumption of multivariate normality. Maximization of $\mathbf{a}'\mathbf{Sa}$ is equivalent to minimization of $\mathbf{a}'\mathbf{S}^{-1}\mathbf{a}$, and looking for the 'smallest' ellipsoids in the family $\mathbf{a}'\mathbf{S}^{-1}\mathbf{a} = $ const. which intersect the hypersphere $\mathbf{a}'\mathbf{a} = 1$, will lead to the largest principal axis of the family $\mathbf{a}'\mathbf{S}^{-1}\mathbf{a}$. Thus the PCs define, successively, the principal axes of the ellipsoids $\mathbf{a}'\mathbf{S}^{-1}\mathbf{a} = $ const., regardless of any assumption of multivariate normality. However, without multivariate normality, the ellipsoids lose their interpretation as contours of equal probability.

Further discussion of the geometry of sample PCs, together with connections with other techniques such as principal co-ordinate analysis (see Section 5.2), and special cases such as compositional data (Section 11.3), is given by Gower (1967).

3.3. Covariance and Correlation Matrices: An Example

The arguments for and against using sample correlation matrices as opposed to covariance matrices are virtually identical to those given for populations in Section 2.3. Furthermore, it is still the case that there is no straightforward relationship between the PCs obtained from a correlation matrix and those based on the corresponding covariance matrix. The main purpose of the present section is to give an example illustrating some of the properties of PCs based on sample covariance and correlation matrices.

The data for this example consist of measurements on eight blood chemistry variables, for 72 patients in a clinical trial. The correlation matrix for these data, together with the standard deviations of each of the eight variables, is given in Table 3.1. Two main points emerge from Table 3.1: first, there are considerable differences in the standard deviations, caused mainly by differences in scale for the eight variables; and, second, none of the correlations is particularly large in absolute value, apart from the value of -0.877 for NEUT and LYMPH.

The large differences in standard deviations give a warning that there may be considerable differences between the PCs for the correlation and covariance matrices. That this is indeed so can be seen in Tables 3.2 and 3.3, which give coefficients for the first four components, based on the correlation and covariance matrix respectively. For ease of comparison, the coefficients are rounded to the nearest 0.2. The effect of such severe rounding is investigated, for this example, in Section 12.4.

Each of the first four PCs for the correlation matrix has moderate-sized coefficients for several of the variables, whereas the first four PCs for the covariance matrix are each dominated by a single variable. The first component is a slight perturbation of the single variable, PLATE, which has the largest variance, the second component is almost the same as the variable, BILIR, with the second highest variance, and so on. In fact, this pattern continues for the fifth and sixth components which are not shown in Table 3.3. Also, the relative percentages of total variation accounted for by each component mirror closely the variances of the corresponding variables. Since variable PLATE has a variance 100 times larger than any other variable, the first PC accounts for over 98% of the total variation. Thus the first six components for the covariance matrix tell us almost nothing apart from the order of sizes of variances of the original variables. By contrast, the first few PCs for the correlation matrix show that certain non-trivial linear functions of the (standardized) original variables account for substantial, though not enormous, proportions of the total variation in the standardized variables. In particular, a weighted contrast between the first four and the last four variables is the linear function with the largest variance.

This example illustrates the dangers in using a covariance matrix to find PCs when the variables have widely differing variances; the first few PCs will contain little information apart from the relative sizes of variances, information which was available without a PCA.

Apart from the fact, already mentioned in Section 2.3, that it is more difficult to base statistical inference regarding PCs on correlation matrices, one other disadvantage of correlation matrix PCs is that they give coefficients for standardized variables and are therefore less easy to interpret directly. To interpret the PCs in terms of the original variables each coefficient must be divided by the standard deviation of the corresponding variable. An example which illustrates this is given in the next section. It must not be forgotten, however, when re-expressing correlation matrix PCs in terms of the original variables, that they are still linear functions of $\mathbf{x}$ which maximize variance

Table 3.1. Correlations and standard deviations for eight blood chemistry variables.

	RBLOOD	PLATE	WBLOOD	NEUT	LYMPH	BILIR	SODIUM	POTASS
				Correlation matrix				
RBLOOD	1.000							
PLATE	0.290	1.000						
WBLOOD	0.202	0.415	1.000					
NEUT	−0.055	0.285	0.419	1.000				
LYMPH	−0.105	−0.376	−0.521	−0.877	1.000			
BILIR	−0.252	−0.349	−0.441	−0.076	0.206	1.000		
SODIUM	−0.229	−0.164	−0.145	0.023	0.034	0.192	1.000	
POTASS	0.058	−0.129	−0.076	−0.131	0.151	0.077	0.423	1.000
Standard deviations	0.371	41.253	1.935	0.077	0.071	4.037	2.732	0.297

Table 3.2. Principal components based on the correlation matrix for eight blood chemistry variables.

Component number	1	2	3	4
	Coefficients			
RBLOOD	0.2	−0.4	0.4	0.6
PLATE	0.4	−0.2	0.2	0.0
WBLOOD	0.4	0.0	0.2	−0.2
NEUT	0.4	0.4	−0.2	0.2
LYMPH	−0.4	−0.4	0.0	−0.2
BILIR	−0.4	0.4	−0.2	0.6
SODIUM	−0.2	0.6	0.4	−0.2
POTASS	−0.2	0.2	0.8	0.0
Percentage of total variation explained	34.9	19.1	15.6	9.7

with respect to the standardized variables and not with respect to the original variables.

An alternative to finding PCs for either covariance or correlation matrices is to calculate the eigenvectors of $\tilde{\mathbf{X}}'\tilde{\mathbf{X}}$ rather than $\mathbf{X}'\mathbf{X}$, i.e. measure variables about zero, rather than about their sample means, when computing 'covariances' and 'correlations'. This idea has been noted by Jöreskog et al. (1976, p. 124) and will be discussed further in Section 12.3. 'Principal component analysis' based on measures of association of this form, but for observations rather than variables, has been found useful for certain types of geological data (Jöreskog et al., 1976, Section 5.2). Yet another variant, in a way the opposite of that just mentioned, has been used by Buckland and Anderson (1985). Their idea, which is also discussed further in Section 12.3, and which again seems appropriate for a particular type of data, is to 'correct for the mean' in both the rows and columns of $\tilde{\mathbf{X}}$.

Table 3.3. Principal components based on the covariance matrix for eight blood chemistry variables.

Component number	1	2	3	4
	Coefficients			
RBLOOD	0.0	0.0	0.0	0.0
PLATE	1.0	0.0	0.0	0.0
WBLOOD	0.0	−0.2	0.0	1.0
NEUT	0.0	0.0	0.0	0.0
LYMPH	0.0	0.0	0.0	0.0
BILIR	0.0	1.0	−0.2	0.2
SODIUM	0.0	0.2	1.0	0.0
POTASS	0.0	0.0	0.0	0.0
Percentage of total variation explained	98.6	0.9	0.4	0.2

3.4. Principal Components with Equal and/or Zero Variances

The problems which arise when some of the eigenvalues of a population covariance matrix are zero and/or equal were discussed in Section 2.4, and similar considerations hold when dealing with a sample.

For a sample covariance or correlation matrix, the occurrence of q zero eigenvalues implies that the points $x_1, x_2, \ldots, x_n$ lie in a $(p - q)$-dimensional subspace of p-dimensional space. This means that there are q separate linear functions of the p original variables which have constant values for each of the observations $x_1, x_2, \ldots, x_n$. Ideally, constant relationships between the variables should be detected before doing a PCA, and the number of variables reduced so as to avoid them. However, prior detection will not always be possible and the zero-variance PCs will enable any unsuspected constant relationships to be detected. Similarly, PCs with very small, but non-zero, variances will define near-constant linear relationships, and finding such near-constant relationships may be of considerable interest. In addition, low-variance PCs have a number of more specific potential uses, as will be discussed at the end of Section 3.7 and in Sections 6.3, 8.4, 8.6 and 10.1.

3.4.1. Example

Here we consider another set of blood chemistry data, this time consisting of 16 variables measured on 36 patients. In fact, these observations and those discussed in the previous section are both subsets of a larger data set. In the present subset, four of the variables, x_1, x_2, x_3, x_4, say, sum to 1.00 for 35 patients and to 0.99 for the remaining patient, so that $x_1 + x_2 + x_3 + x_4$ is nearly constant. The last (sixteenth) PC for the correlation matrix has a very small variance (less than 0.001), much smaller than the fifteenth, and is (rounding coefficients to the nearest 0.1) $0.7x_1^* + 0.3x_2^* + 0.7x_3^* + 0.1x_4^*$, with all of the other 12 variables having negligible coefficients. Thus, the near-constant relationship has certainly been identified by the last PC, but not in an easily interpretable form. However, a simple interpretation can be restored if the standardized variables are replaced by the original variables, by setting $x_j^* = x_j/s_{jj}^{1/2}$, where $s_{jj}^{1/2}$ is the sample standard deviation of x_j. When this is done, the last PC becomes (rounding coefficients to the nearest integer), $11x_1 + 11x_2 + 11x_3 + 11x_4$. The correct near-constant relationship has therefore been discovered exactly (to the degree of rounding used) by the last PC.

With carefully selected variables, PCs with zero variances are a relatively rare occurrence. Equal variances for sample PCs are even more unusual; even if the underlying population covariance (correlation) matrix has a pattern which gives equal eigenvalues, sampling variation will usually ensure that no two sample PCs have exactly the same variance.

3.5. The Singular Value Decomposition

This section describes a result from matrix theory, namely the singular value decomposition (SVD), which is relevant to PCA in several respects. Given an arbitrary matrix, X, of dimension $n \times p$, which for our purposes will invariably be a matrix of n observations on p variables, measured about their means, X can be written

$$X = ULA', \qquad (3.5.1)$$

where

(i) U, A are $(n \times r)$, $(p \times r)$ matrices respectively, each of which has ortho-normal columns so that $U'U = I_r$, $A'A = I_r$;
(ii) L is a $(r \times r)$ diagonal matrix;
(iii) r is the rank of X.

To prove this result, consider the spectral decomposition of $X'X$. The last $p - r$ terms in (3.1.4), and in the corresponding expression for $X'X$, are zero, since the last $p - r$ eigenvalues are zero if X, and hence $X'X$, has rank r. Thus

$$(n - 1)S = X'X = l_1 a_1 a_1' + l_2 a_2 a_2' + \cdots + l_r a_r a_r'.$$

[Note that in this section it is convenient to denote the eigenvalues of $X'X$, rather than those of S, as l_k, $k = 1, 2, \ldots, p$.] Define A to be the $(p \times r)$ matrix with kth column a_k, define U as the $(n \times r)$ matrix whose kth column is

$$u_k = l_k^{-1/2} X a_k, \qquad k = 1, 2, \ldots, r,$$

and define L to be the $(r \times r)$ diagonal matrix with kth diagonal element $l_k^{1/2}$. Then U, L, A satisfy conditions (i) and (ii) above, and we shall now show that $X = ULA'$.

$$ULA' = U \begin{bmatrix} l_1^{1/2} a_1' \\ l_2^{1/2} a_2' \\ \vdots \\ l_r^{1/2} a_r' \end{bmatrix}$$

$$= \sum_{k=1}^{r} l_k^{-1/2} X a_k l_k^{1/2} a_k' = \sum_{k=1}^{r} X a_k a_k'$$

$$= \sum_{k=1}^{p} X a_k a_k',$$

because a_k, $k = r + 1, r + 2, \ldots, p$ are eigenvectors of $X'X$ corresponding to zero eigenvalues, so that $X'X a_k = 0$ and hence $X a_k = 0$, $k = r + 1, r + 2, \ldots, p$. Thus

$$ULA' = X \sum_{k=1}^{p} a_k a_k' = X,$$

as required, because the $(p \times p)$ matrix whose kth column is a_k, is orthogonal, and so has orthonormal rows.

✴ The importance of the SVD for PCA is twofold. First, it provides a computationally efficient method of actually finding PCs—see the first section of the Appendix. It is clear that if we can find **U**, **L**, **A** which satisfy (3.5.1), then **A** and **L** will give us the eigenvectors and the square roots of the eigenvalues of **X′X**, and hence the principal components and their variances for the sample covariance matrix, **S**. As a bonus we also get, in **U**, standardized versions of PC scores. To see this multiply (3.5.1) on the right by **A** to give **XA** = **ULA′A** = **UL** since **A′A** = **I**$_r$. But **XA** is an $(n \times r)$ matrix whose kth column consists of the PC scores for the kth PC—see (3.1.2) for the case where $r = p$. The PC scores z_{ik} are, therefore, given by

$$z_{ik} = u_{ik} l_k^{1/2}, \qquad i = 1, 2, \ldots, n, \quad k = 1, 2, \ldots, r,$$

or, in matrix form, **Z** = **UL**, or **U** = **ZL**$^{-1}$. The variance of the scores for the kth PC is $l_k/(n-1)$, $k = 1, 2, \ldots, p$. [Recall that l_k, here, denotes the kth eigenvalue of **X′X**, so that the kth eigenvalue of **S** is $l_k/(n-1)$.] Therefore the scores given by **U** are simply those given by **Z**, but standardized to have variance $1/(n-1)$. Note also that the columns of **U** are the eigenvectors of **XX′** corresponding to non-zero eigenvalues, and these eigenvectors are of potential interest if the rôles of 'variables' and 'observations' are reversed.

The second virtue of the SVD is that it provides additional insight into what a PCA actually does, and gives useful means, both graphical and algebraic, of representing the results of a PCA. This has been recognized in different contexts by Mandel (1972), Gabriel (1978), Rasmusson et al. (1981) and Eastment and Krzanowski (1982), and will be discussed further in connection with relevant applications in Sections 5.3, 6.1.5, 11.4, 11.5 and 11.6. Furthermore, the SVD is useful in terms of both computation and interpretation in PC regression—see Section 8.1 and Mandel (1982)—and in examining the links between PCA and correspondence analysis (Sections 11.1 and 12.1).

In the meantime, note that (3.5.1) can be written element by element as

$$x_{ij} = \sum_{k=1}^{r} u_{ik} l_k^{1/2} a_{jk}, \qquad (3.5.2)$$

where u_{ik}, a_{jk} are the (i, k)th, (j, k)th elements of **U**, **A** respectively and $l_k^{1/2}$ is the kth element of **L**. Thus x_{ij} can be split into parts

$$u_{ik} l_k^{1/2} a_{jk}, \qquad k = 1, 2, \ldots, r,$$

corresponding to each of the first r PCs, and if only the first m PCs are retained, then

$$_m \tilde{x}_{ij} = \sum_{k=1}^{m} u_{ik} l_k^{1/2} a_{jk} \qquad (3.5.3)$$

✓will provide an approximation to x_{ij}. In fact, it can be shown (Gabriel, 1978; Householder and Young, 1938) that $_m \tilde{x}_{ij}$ provides the best possible rank m approximation to x_{ij} in the sense of minimizing

$$\sum_{i=1}^{n} \sum_{j=1}^{p} (_m x_{ij} - x_{ij})^2,$$

where $_m x_{ij}$ is any rank m approximation to x_{ij}. Another way of expressing this result is that the $(n \times p)$ matrix whose (i, j)th element is $_m \tilde{x}_{ij}$ minimizes $\| _m\mathbf{X} - \mathbf{X} \|$ over all $(n \times p)$ matrices, $_m\mathbf{X}$, with rank m. Thus the SVD provides a sequence of approximations, of rank $1, 2, \ldots, r$, to $\mathbf{X}$, which minimize the Euclidean norm of the difference between $\mathbf{X}$ and the approximation $_m\mathbf{X}$. This result provides an interesting parallel to the result given earlier (see the proof of Property G4 in Section 3.2) that the spectral decomposition of $\mathbf{X}'\mathbf{X}$ provides a similar optimal sequence of approximations of rank $1, 2, \ldots, r$, to the matrix $\mathbf{X}'\mathbf{X}$.

3.6. Probability Distributions for Sample Principal Components

A considerable amount of mathematical effort has been expended on deriving probability distributions, mostly asymptotic, for the coefficients in the sample PCs, and for the variances of sample PCs or, equivalently, finding distributions for the eigenvectors and eigenvalues of a sample covariance matrix. For example, the first issue of the *Journal of Multivariate Analysis* in 1982 contained three papers, totalling 83 pages, on the topic. The distributional results provided in these papers and elsewhere have three drawbacks:

(i) they usually involve complicated mathematics;
(ii) they are mostly asymptotic;
(iii) they are often based on the assumption that the original set of variables has a multivariate normal distribution.

Despite these drawbacks, some of the main results are given below, and their use in inference about the population PCs, given sample PCs, is discussed in the next section.

Assume that $\mathbf{x} \sim N(\boldsymbol{\mu}, \boldsymbol{\Sigma})$, that is $\mathbf{x}$ has a p-variate normal distribution with mean $\boldsymbol{\mu}$, and covariance matrix $\boldsymbol{\Sigma}$. Although $\boldsymbol{\mu}$ need not be given, $\boldsymbol{\Sigma}$ is assumed known. Then

$$(n - 1)\mathbf{S} \sim W_p(\boldsymbol{\Sigma}, n - 1),$$

i.e. $(n - 1)\mathbf{S}$ has the so-called Wishart distribution with parameters $\boldsymbol{\Sigma}, n - 1$ (see, for example, Mardia *et al.*, 1979, Section 3.4). Therefore, investigation of the sampling properties of the coefficients and variances of the sample PCs is equivalent to looking at sampling properties of eigenvectors and eigenvalues of Wishart random variables.

The density function of a matrix $\mathbf{V}$ which has the $W_p(\boldsymbol{\Sigma}, n - 1)$ distribution is

$$c|\mathbf{V}|^{(n-p-2)/2}\exp\{-\tfrac{1}{2}\operatorname{tr}(\boldsymbol{\Sigma}^{-1}\mathbf{V})\},$$

where

$$c^{-1}=2^{p(n-1)/2}\,\Pi^{p(1-p)/4}|\boldsymbol{\Sigma}|^{(n-1)/2}\prod_{j=1}^{p}\Gamma\left(\frac{n-j}{2}\right),$$

and various properties of Wishart random variables have been thoroughly investigated (see, for example, Srivastava and Khatri, 1979, Chapter 3).

Let l_k, $\mathbf{a}_k$, $k = 1, 2, \ldots, p$ be the eigenvalues and eigenvectors of $\mathbf{S}$, and let λ_k, $\boldsymbol{\alpha}_k$, $k = 1, 2, \ldots, p$ be the eigenvalues and eigenvectors of $\boldsymbol{\Sigma}$. Also, let $\mathbf{l}$, $\boldsymbol{\lambda}$ be the p-element vectors consisting of the l_k's and λ_k's respectively and let the jth elements of $\mathbf{a}_k$, $\boldsymbol{\alpha}_k$ be a_{kj}, α_{kj} respectively. [The notation a_{jk} was used for the jth element of $\mathbf{a}_k$ in the previous section, but it seems more natural to use a_{kj} in this, and the next, section.] The best known and simplest results concerning the distribution of the l_k's and the $\mathbf{a}_k$'s assume, usually quite realistically, that $\lambda_1 > \lambda_2 > \cdots > \lambda_p > 0$, i.e. all the population eigenvalues are positive and distinct. Then the following results hold *asymptotically*:

(i) all of the l_k's are independent of all of the $\mathbf{a}_k$'s;
(ii) $\mathbf{l}$ and the $\mathbf{a}_k$'s are jointly normally distributed;
(iii) $E(\mathbf{l}) = \boldsymbol{\lambda}$; $E(\mathbf{a}_k) = \boldsymbol{\alpha}_k$, $k = 1, 2, \ldots, p$;   (3.6.1)

(iv) $\operatorname{Cov}(l_k, l_{k'}) = \begin{cases} 2\lambda_k^2/(n-1), & k = k', \\ 0, & k \neq k', \end{cases}$ (3.6.2)

$$\operatorname{Cov}(a_{kj}, a_{k'j'}) = \begin{cases} \dfrac{\lambda_k}{(n-1)}\displaystyle\sum_{\substack{l=1 \\ l \neq k}}^{p}\dfrac{\lambda_l \alpha_{lj}\alpha_{lj'}}{(\lambda_l - \lambda_k)^2}, & k = k', \\[4mm] -\dfrac{\lambda_k \lambda_{k'}\alpha_{kj}\alpha_{k'j'}}{(n-1)(\lambda_k - \lambda_{k'})^2} & k \neq k'. \end{cases}$$ (3.6.3)

An extension of the above results to the case where some of the λ_k may be equal to each other, though still positive, was given by Anderson (1963) and an alternative proof can be found in Srivastava and Khatri (1979, Section 9.4.1).

It should be stressed that the above results are asymptotic and therefore only approximate for finite samples. Exact results are available, but only for a few special cases, such as when $\boldsymbol{\Sigma} = \mathbf{I}$ (Srivastava and Khatri, 1979, p. 86) and more generally for l_1, l_p, the largest and smallest eigenvalues (Srivastava and Khatri, 1979, p. 205). In addition, better but more complicated approximations can be found to the distributions of $\mathbf{l}$ and the $\mathbf{a}_k$'s in the general case (Srivastava and Khatri, 1979, Section 9.4, and the references cited there).

If a distribution other than the multivariate normal is assumed, distributional results for PCs will typically become less tractable. In addition, for non-normal distributions a number of alternatives to PCs can reasonably be suggested—see Sections 11.1, 11.3 and 12.6.

As a complete contrast to the strict assumptions made in most work on the distributions of PCs, Diaconis and Efron (1983) look at the use of the 'bootstrap' in this context. The idea is, for a particular sample of n observations $\mathbf{x}_1, \mathbf{x}_2, \ldots, \mathbf{x}_n$, to take repeated random samples of size n from the distribution which has $P[\mathbf{x} = \mathbf{x}_i] = 1/n$, $i = 1, 2, \ldots, n$, calculate the PCs for each sample, and build up empirical distributions for PC coefficients and variances. These distributions rely only the structure of the sample, and not on any predetermined assumptions.

3.7. Inference Based on Sample Principal Components

The distributional results outlined in the previous section may be used to make inferences about the population PCs, given the sample PCs, provided that the necessary assumptions are valid. The major assumption that $\mathbf{x}$ has a multivariate normal distribution, is often not satisfied, and the practical value of the results is therefore limited. It can be argued that PCA should only ever be done for data which is, at least approximately, multivariate normal, for it is only then that 'proper' inferences can be made regarding the underlying population PCs. However, this seems a rather narrow view of what PCA can do, since it is, in fact, a much more widely applicable tool, whose main use is descriptive rather than inferential. It can provide valuable descriptive information for a wide variety of data, whether the variables are continuous and normally distributed or not. The majority of applications of PCA successfully treat the technique as a purely descriptive tool, although Mandel (1972) has argued that retaining m PCs in an analysis implicitly assumes a model for the data, based on (3.5.3).

The purely inferential side of PCA is really a very small part of the overall picture, based as it is on restrictive assumptions and on (mostly) asymptotic results. However, the ideas of inference can sometimes be useful and are discussed briefly in the next three subsections. One of the few published examples, which illustrate the use of equations (3.6.2) and (3.6.3) on a real data set, is given by Jackson and Hearne (1973). This article also gives a good discussion of the practical implications of these equations.

3.7.1. Point Estimation

The maximum likelihood estimator (MLE) for $\boldsymbol{\Sigma}$, the covariance matrix of a multivariate normal distribution, is not $\mathbf{S}$, but $(n - 1)\mathbf{S}/n$ (see, for example, Press, 1972, Section 7.1 for a derivation). This result is hardly surprising, given the corresponding result for the univariate normal. If $\lambda, \mathbf{l}, \boldsymbol{\alpha}_k, \mathbf{a}_k$ and related quantities are defined as in the previous section, then the MLEs of λ and $\boldsymbol{\alpha}_k$, $k = 1, 2, \ldots, p$, can be derived from the MLE of $\boldsymbol{\Sigma}$ and are equal to $\hat{\lambda} = (n - 1)\mathbf{l}/n$, and $\hat{\boldsymbol{\alpha}}_k = \mathbf{a}_k$, $k = 1, 2, \ldots, p$, assuming that the elements of λ

are all positive and distinct. These MLEs are the same, in this case, as the estimators derived by the method of moments. The MLE for λ_k is biased but asymptotically unbiased, as is the MLE for **S**.

In the case where some of the λ_k's are equal, the MLE for their common value is simply the average of the corresponding l_k's, multiplied by $(n - 1)/n$. The MLEs of the $\boldsymbol{\alpha}_k$'s corresponding to equal λ_k's are not unique, in exactly the same way as the corresponding $\mathbf{a}_k$'s are not unique; the $(p \times q)$ matrix whose columns are MLEs of $\boldsymbol{\alpha}_k$'s corresponding to equal λ_k's can be multiplied by any $(q \times q)$ orthogonal matrix, where q is the multiplicity of the eigenvalues, to get another set of MLEs.

If multivariate normality cannot be assumed, and if there is no obvious alternative distributional assumption, then it may be desirable to use a 'robust' approach to the estimation of the PCs; this topic is discussed in Section 10.3.

3.7.2. Interval Estimation

The asymptotic marginal distributions of l_k and a_{kj} given in the previous section can be used to construct approximate confidence intervals for λ_k and α_{kj} respectively. For l_k, the marginal distribution is, from (3.6.1) and (3.6.2), approximately

$$l_k \sim N(\lambda_k, 2\lambda_k^2/(n - 1)), \tag{3.7.1}$$

so

$$\frac{l_k - \lambda_k}{\lambda_k[2/(n - 1)]^{1/2}} \sim N(0, 1),$$

which leads to a confidence interval, with confidence coefficient $1 - \alpha$, for λ_k, of the form

$$\frac{l_k}{[1 + \tau z_{\alpha/2}]^{1/2}} < \lambda_k < \frac{l_k}{[1 - \tau z_{\alpha/2}]^{1/2}}, \tag{3.7.2}$$

where $\tau^2 = 2/(n - 1)$, and $z_{\alpha/2}$ is the upper $(100)\alpha/2$ percentile of the standard normal distribution, $N(0, 1)$. In deriving this confidence interval it is assumed that n is large enough so that $\tau z_{\alpha/2} < 1$. Since the distributional result is asymptotic, this is a realistic assumption. An alternative approximate confidence interval is obtained by looking at the distribution of $\log_e l_k$. Given (3.7.1) it follows that

$$\log_e l_k \sim N(\log_e \lambda_k, 2/(n - 1)) \quad \text{approximately,}$$

thus removing the dependence of the variance on the unknown parameter λ_k. An approximate confidence interval for $\log_e \lambda_k$, with confidence coefficient $1 - \alpha$, is then $\log_e l_k \pm \tau z_{\alpha/2}$, and transforming back to λ_k gives an approximate confidence interval of the form

$$l_k \, e^{-\tau z_{\alpha/2}} \leq \lambda_k \leq l_k \, e^{\tau z_{\alpha/2}}. \qquad (3.7.3)$$

Since the l_k's are independent, joint confidence regions for several of the λ_k's are obtained by simply combining intervals of the form (3.7.2) or (3.7.3), choosing individual confidence coefficients so as to achieve an overall desired confidence level. Approximate confidence intervals for individual α_{kj}'s can be obtained from the marginal distributions of the a_{kj}'s whose means and variances are given in (3.6.1) and (3.6.3). The intervals are constructed in a similar manner to those for the λ_k's, although the expressions involved are somewhat more complicated. Expressions become still more complicated when looking at joint confidence regions for several α_{kj}'s, partly because of the non-independence of separate a_{kj}'s. Consider $\mathbf{a}_k$: from (3.6.1), (3.6.3) it follows that, approximately,

$$\mathbf{a}_k \sim N(\boldsymbol{\alpha}_k, \mathbf{T}_k),$$

where

$$\mathbf{T}_k = \frac{\lambda_k}{(n-1)} \sum_{\substack{l=1 \\ l \neq k}}^{p} \frac{\lambda_l}{(\lambda_l - \lambda_k)^2} \boldsymbol{\alpha}_l \boldsymbol{\alpha}_l'.$$

$\mathbf{T}_k$ has rank $(p-1)$, since it has one zero eigenvalue corresponding to the eigenvector $\boldsymbol{\alpha}_k$, which causes further complications, but it can be shown that (Mardia *et al.*, 1979, p. 233), approximately,

$$(n-1)(\mathbf{a}_k - \boldsymbol{\alpha}_k)'(l_k \mathbf{S}^{-1} + l_k^{-1}\mathbf{S} - 2\mathbf{I}_p)(\mathbf{a}_k - \boldsymbol{\alpha}_k) \sim \chi^2_{(p-1)}, \qquad (3.7.4)$$

Because $\mathbf{a}_k$ is an eigenvector of $\mathbf{S}$ with eigenvalue l_k, it follows that $l_k^{-1}\mathbf{S}\mathbf{a}_k = l_k^{-1}l_k\mathbf{a}_k = \mathbf{a}_k$, $l_k\mathbf{S}^{-1}\mathbf{a}_k = l_k l_k^{-1}\mathbf{a}_k = \mathbf{a}_k$, and $(l_k\mathbf{S}^{-1} + l_k^{-1}\mathbf{S} - 2\mathbf{I}_p)\mathbf{a}_k = \mathbf{a}_k + \mathbf{a}_k - 2\mathbf{a}_k = \mathbf{0}$, so that the result (3.7.4) reduces to

$$(n-1)\boldsymbol{\alpha}_k'(l_k\mathbf{S}^{-1} + l_k^{-1}\mathbf{S} - 2\mathbf{I}_p)\boldsymbol{\alpha}_k \sim \chi^2_{(p-1)}. \qquad (3.7.5)$$

From (3.7.5) an approximate confidence region for $\boldsymbol{\alpha}_k$, with confidence coefficient $1 - \alpha$, has the form $(n-1)\boldsymbol{\alpha}_k'(l_k\mathbf{S}^{-1} + l_k^{-1}\mathbf{S} - 2\mathbf{I}_p)\boldsymbol{\alpha}_k \leq \chi^2_{(p-1);\alpha}$, with fairly obvious notation.

3.7.3. Hypothesis Testing

The same results, obtained from (3.6.1)–(3.6.3), which were used above to derive confidence intervals for individual l_k's and a_{kj}'s are also useful for constructing tests of hypotheses. For example, if it is required to test $H_0: \lambda_k = \lambda_{k0}$ against $H_1: \lambda_k \neq \lambda_{k0}$, then a suitable test statistic is

$$\frac{l_k - \lambda_{k0}}{\tau \lambda_{k0}},$$

which has, approximately, a $N(0, 1)$ distribution under H_0, so that H_0 would be rejected, at significance level α, if

$$\left| \frac{l_k - \lambda_{k0}}{\tau \lambda_{k0}} \right| \geq z_{\alpha/2}.$$

Similarly, the result (3.7.5) can be used to test $H_0: \alpha_k = \alpha_{k0}$ against $H_1: \alpha_k \neq \alpha_{k0}$. A test of H_0 against H_1 will reject H_0, at significance level α, if

$$(n - 1)\alpha'_{k0}(l_k S^{-1} + l_k^{-1} S - 2I_p)\alpha_{k0} \geq \chi^2_{(p-1);\alpha}.$$

This is, of course, an approximate test, but some exact tests, assuming multivariate normality of x, are also available (Srivastava and Khatri, 1979, Section 9.7). Details will not be given here, partly because it is very unusual that a particular pattern can be postulated for the coefficients of an individual population PC, so that such tests are of limited practical use.

There are, however, tests concerning other types of patterns in Σ and its eigenvalues and eigenvectors which are perhaps more useful. The best known of these is the test of $H_{0q}: \lambda_{q+1} = \lambda_{q+2} = \cdots = \lambda_p$ (i.e. the last $p - q$ eigenvalues are equal) against the alternative, H_1, that at least two of the last $p - q$ eigenvalues are different. In his original paper, Hotelling (1933) looked at the problem of testing the equality of *two* consecutive eigenvalues, and tests of H_{0q} have since been considered by a number of authors, including Bartlett (1950), whose name is sometimes given to such tests. The justification for wishing to test H_{0q} is that the first q PCs may each be measuring some substantial component of variation in x, but the last $(p - q)$ PCs are of equal variation and essentially just measure 'noise'. Geometrically, this means that the distribution of the last $(p - q)$ PCs has spherical contours of equal probability, assuming multivariate normality, and the last $(p - q)$ PCs are therefore not uniquely defined. By testing H_{0q} for various values of q, it can be decided how many PCs are distinguishable from 'noise' and are therefore worth retaining. This idea for deciding how many components to retain will be discussed critically in Section 6.1.4.

A test statistic for H_{0q} against a general alternative, H_{1q}, that at least two of the last $(p - q)$ λ_k's are unequal, can be found by assuming multivariate normality, and constructing a likelihood ratio (LR) test. The test statistic takes the form

$$Q = \left\{ \prod_{k=q+1}^{p} l_k \middle/ \left[\sum_{k=q+1}^{p} l_k/(p - q) \right]^{p-q} \right\}^{n/2}.$$

The exact distribution of Q is complicated, but we can use the well-known general result from statistical inference concerning LR tests, namely that $-2 \log_e Q$ has, approximately, a χ^2 distribution with degrees of freedom equal to the difference between the number of independently varying parameters under $H_{0q} \cup H_{1q}$ and under H_{0q}. Calculating the number of degrees of freedom is non-trivial (Mardia *et al.*, 1979, p. 235) but it turns out to be $\nu = \frac{1}{2}(p - q + 2)(p - q - 1)$, so approximately, under H_{0q}

$$n\left[(p-q)\log_e \bar{l} - \sum_{k=q+1}^{p} \log_e l_k\right] \sim \chi^2_{(v)}, \tag{3.7.6}$$

where

$$\bar{l} = \sum_{k=q+1}^{p} l_k/(p-q).$$

In fact, the approximation can be improved if n is replaced by $n' = n - (2p + 11)/6$, so H_{0q} is rejected, at significance level α, if

$$n'\left[(p-q)\log_e \bar{l} - \sum_{k=q+1}^{p} \log_e l_k\right] \geq \chi^2_{v;\alpha}.$$

[Another, more complicated, improvement to the approximation is given in Srivastava and Khatri, 1979, p. 294.]

A special case of this test occurs when $q = 0$, in which case H_{0q} is equivalent to all the variables being independent and having equal variances, a very restrictive assumption. The test with $q = 0$ would reduce to a test that all variables were independent, with no requirement of equal variances, if we were dealing with a correlation matrix. However, it should be noted that all the results in this and the previous section are for covariance, not correlation, matrices which restricts their usefulness still further.

In general, inference concerning PCs of correlation matrices is more complicated than that for covariance matrices (Anderson, 1963), since the off-diagonal elements of a correlation matrix are non-trivial functions of the random variables which make up the elements of a covariance matrix. For example, the asymptotic distribution of the test statistic (3.7.6) is no longer χ^2, for the correlation matrix, although Lawley (1963) provides an alternative statistic, for a special case, which does have a limiting χ^2 distribution.

Another special case of the test based on (3.7.6) occurs when it is required to test

$$H_0: \Sigma = \sigma^2 \begin{bmatrix} 1 & \rho & \cdots & \rho \\ \rho & 1 & \cdots & \rho \\ \vdots & \vdots & & \vdots \\ \rho & \rho & \cdots & 1 \end{bmatrix}$$

against a general alternative. H_0 states that all variables have the same variance σ^2, and all pairs of variables have the same correlation ρ, in which case

$$\sigma^2[1 + (p-1)\rho] = \lambda_1 > \lambda_2 = \lambda_3 = \cdots = \lambda_p = \sigma^2(1-\rho)$$

(Morrison, 1976, Section 8.6), so that the last $(p-1)$ eigenvalues are equal. If ρ, σ^2 are unknown, then the earlier test is appropriate with $q = 1$, but if ρ, σ^2 are specified then a different test can be constructed, again based on the LR criterion.

Further tests regarding λ and the $\mathbf{a}_k$'s can be constructed, such as the test, discussed by Mardia *et al.* (1979, Section 8.4.2) that the first q PCs account for a given proportion of the total variation. However, as stated at the beginning of this section, these tests are of limited value in practice. Not only are most of the tests asymptotic and/or approximate, but they also rely on the assumption of multivariate normality. Furthermore, it is arguable whether it is often possible to formulate a particular hypothesis which it is of interest to test. More usually, PCA is used to explore the data, rather than to verify predetermined hypotheses.

To conclude this section on inference, we outline a possible use of PCA when a Bayesian approach to inference is adopted. Suppose that $\boldsymbol{\theta}$ is a vector of parameters, and that the posterior distribution for $\boldsymbol{\theta}$ has covariance matrix Σ. If we find PCs for $\boldsymbol{\theta}$, then the last few PCs will provide information on which linear functions of the elements of $\boldsymbol{\theta}$ can be estimated with high precision (low variance). Conversely, the first few PCs are linear functions of the elements of $\boldsymbol{\theta}$ which can only be estimated with low precision. In this context, then, it would seem that the last few PCs may be more useful than the first few.

3.8. Principal Components for Patterned Correlation or Covariance Matrices

At the end of Chapter 2, and in Section 3.7.3, the structure of the PCs, and their variances, was discussed briefly for a correlation matrix with equal correlations between all variables. Other theoretical patterns in correlation and covariance matrices can also be investigated—for example, Jolliffe (1970) considers correlation matrices, with elements ρ_{ij}, for which

$$\rho_{1j} = \rho, \qquad j = 2, 3, \ldots, p,$$

and

$$\rho_{ij} = \rho^2, \qquad 2 \le i < j < p,$$

and Brillinger (1981, p. 108) discusses PCs for Toeplitz matrices, which occur for time series data (see Section 11.2), and in which ρ_{ij} depends only on $|i - j|$.

Such exact patterns will not, in general, occur in sample covariance or correlation matrices, but it is often possible to deduce the approximate form of some of the PCs by recognizing a particular type of structure in a sample covariance or correlation matrix. One such pattern, which was discussed in Section 3.3, occurs when one or more of the variances in a covariance matrix are of very different sizes from all the rest. In this case, as illustrated in the example of Section 3.3, there will be a PC associated with each such variable, which is almost indistinguishable from that variable. Similar behaviour (i.e. a PC very similar to one of the original variables) can occur for correlation matrices, but in rather different circumstances. Here the criterion for such a

PC is that the corresponding variable is nearly uncorrelated with all of the other variables.

The other main type of pattern which can be detected in many correlation matrices is that there are one or more groups of variables, within which all correlations are positive and not close to zero. (Sometimes a variable in such a group will initially have entirely negative correlations with the other members of the group, but the sign of a variable is usually arbitrary, and switching the sign will give a group of the required structure.) If correlations between the q members of the group and variables outside the group are close to zero, then there will be q PCs 'associated with the group' whose coefficients for variables outside the group are small. One of these PCs will have a large variance, approximately $1 + (q - 1)\bar{r}$, where $\bar{r}$ is the average correlation within the group, and will have positive coefficients for all variables in the group. The remaining $(q - 1)$ PCs will have much smaller variances (of order, $1 - \bar{r}$), and will have some positive and some negative coefficients. Thus the 'large variance PC' for the group measures, roughly, the average size of variables in the group, whereas the 'small variances PCs' give 'contrasts' between some or all of the variables in the group. There may be several such groups of variables in a data set, in which case each group will have one 'large variance PC' and several 'small variance PCs'. Conversely, as happens not infrequently, especially in biology when all variables are measurements on individuals of some species, we may find that all p variables are positively correlated. In such cases, the first PC is often interpreted as a measure of size of the individuals, whereas subsequent PCs measure aspects of shape—see Section 4.1 for further discussion.

The discussion above implies that the approximate structure and variances of the first few PCs can be deduced from a correlation matrix, provided that well-defined groups of variables (including possibly single-variable groups) can be detected, whose within-group correlations are high, and whose between-group correlations are low. The ideas can be taken further, and lower bounds on the variance of the first PC can be calculated, based on averages of correlations (Friedman and Weisberg, 1981). However, it should be stressed that although data sets for which there is some group structure among variables are not uncommon, there are many others for which no such pattern is apparent; in such cases the structure of the PCs cannot usually be found without actually doing the PCA.

3.8.1. Example

In many of the examples discussed in later chapters, it will be seen that the structure of some of the PCs can be partially deduced from the correlation matrix, using the ideas just discussed. Here we describe an example in which *all* the PCs have a fairly clear pattern. The data consist of measurements of strengths of reflexes at ten sites of the body, measured for 143 individuals. As

Table 3.4. Correlation matrix for ten variables measuring reflexes.

	V1	V2	V3	V4	V5	V6	V7	V8	V9	V10
V1	1.00									
V2	0.98	1.00								
V3	0.60	0.62	1.00							
V4	0.71	0.73	0.88	1.00						
V5	0.55	0.57	0.61	0.68	1.00					
V6	0.55	0.57	0.56	0.68	0.97	1.00				
V7	0.38	0.40	0.48	0.53	0.33	0.33	1.00			
V8	0.25	0.28	0.42	0.47	0.27	0.27	0.90	1.00		
V9	0.22	0.21	0.19	0.23	0.16	0.19	0.40	0.41	1.00	
V10	0.20	0.19	0.18	0.21	0.13	0.16	0.39	0.40	0.94	1.00

with the examples discussed in Sections 3.3 and 3.4, the data were kindly supplied by Richard Hews of Pfizer Central Research.

The correlation matrix for these data is given in Table 3.4, and the coefficients of, and the variation explained by, the corresponding PCs are presented in Table 3.5. It should first be noted that the ten variables fall into five pairs. Thus, V1, V2, respectively, denote strength of reflexes for right and left triceps, with {V3, V4}, {V5, V6}, {V7, V8}, {V9, V10} similarly defined for right and left biceps, right and left wrists, right and left knees, and right and left ankles. The correlations between variables within each pair are large, so that the differences between variables in each pair have small variances. This is reflected in the last five PCs, which are mainly within-pair contrasts, with the more highly correlated pairs corresponding to the later components.

Turning to the first two PCs, there is a suggestion in the correlation matrix that, although all correlations are positive, the variables can be divided into two groups {V1–V6}, {V7–V10}. These correspond to sites in the arms and legs respectively. Reflecting this group structure, the first and second PCs have their largest coefficients on the first and second groups of variables, respectively. Because the group structure is not clear-cut, these two PCs also have contributions from the less dominant group, and the first PC is a weighted average of variables from both groups, whereas the second PC is a weighted contrast between the groups.

The third, fourth and fifth PCs reinforce the idea of the two groups. The third PC is a contrast between the two pairs of variables in the second (smaller) group and the fourth and fifth PCs both give contrasts between the three pairs of variables in the first group.

It is relatively rare for examples with as many as ten variables to have such a nicely-defined structure for their PCs as in the present case. However, as will be seen in the examples of subsequent chapters, it is not unusual to be able to deduce the structure of at least a few PCs in this manner.

Table 3.5. Principal components based on the correlation matrix of Table 3.4.

Component number	1	2	3	4	5	6	7	8	9	10
					Coefficients					
V1	0.3	−0.2	0.2	−0.5	0.3	0.1	−0.1	−0.0	−0.6	0.2
V2	0.4	−0.2	0.2	−0.5	0.3	0.0	−0.1	−0.0	0.7	−0.3
V3	0.4	−0.1	−0.1	−0.0	−0.7	0.5	−0.2	0.0	0.1	0.1
V4	0.4	−0.1	−0.1	−0.0	−0.4	−0.7	0.3	−0.0	−0.1	−0.1
V5	0.3	−0.2	0.1	0.5	0.2	0.2	−0.0	−0.1	−0.2	−0.6
V6	0.3	−0.2	0.2	0.5	0.2	−0.1	−0.0	0.1	0.2	0.6
V7	0.3	0.3	−0.5	−0.0	0.2	0.3	0.7	0.0	−0.0	0.0
V8	0.3	0.3	−0.5	0.1	0.2	−0.2	−0.7	−0.0	−0.0	−0.0
V9	0.2	0.5	0.4	0.0	−0.1	0.0	−0.0	0.7	−0.0	−0.1
V10	0.2	0.5	0.4	0.0	−0.1	0.0	0.0	−0.7	0.0	0.0
Percentage of total variation explained	52.3	20.4	11.0	8.5	5.0	1.0	0.9	0.6	0.2	0.2

Principal Components as a Small Number of Interpretable Variables: Some Examples

The original purpose of PCA was to reduce a large number (p) of variables to a much small number (m) of PCs whilst retaining as much as possible of the variation in the p original variables. The technique is especially useful if $m \ll p$, and if the m PCs can be readily interpreted.

Although we shall see in subsequent chapters that there are many other ways of applying PCs, the original usage as a descriptive, dimension-reducing technique is probably still the most prevalent application. This chapter simply introduces a number of examples, from several different fields of application, where PCA not only reduces the dimensionality of the problem substantially, but has PCs which are easily interpreted. Graphical representations of a set of observations with respect to the m retained PCs and discussion of how to choose an appropriate value of m are deferred until Chapters 5 and 6 respectively.

Of course, if m is very much smaller than p, then the reduction of dimensionality alone may justify the use of PCA, even if the PCs have no clear meaning, but the results of a PCA are much more satisfying if intuitively reasonable interpretations can be given to some or all of the m retained PCs.

Each section of this chapter describes one example in detail, but other examples in related areas are also mentioned in most sections. Some of the examples introduced in this chapter are discussed further in subsequent chapters; conversely, when new examples are introduced later in the book, an attempt will be made to interpret the first few PCs, where appropriate. The examples are drawn from a variety of fields of application, demonstrating the fact that PCA has been found useful in a very large number of subject areas, of which those illustrated in this book form only a subset.

It must be emphasized that although in many examples the PCs can be readily interpreted, this is by no means universally true. There is no reason, *a priori*, why a mathematically derived linear function of the original variables (which is what PCs are) should have a simple interpretation. It is remarkable how often it seems to be possible to interpret the first few PCs, though it is probable that some interpretations owe a lot to the analyst's ingenuity and imagination. Careful thought should go into any interpretations and, at an earlier stage, into the choice of variables and whether to transform them. In some circumstances, transformation of variables before analysis may improve the chances of a simple interpretation—see Sections 11.3 and 12.2. Conversely, the arbitrary inclusion of logarithms, powers, ratios, etc., of the original variables can make it unlikely that any simple interpretation will be found.

4.1. Anatomical Measurements

One type of application where PCA has been found useful is in identifying the most important sources of variation in anatomical measurements for various species. Typically, a large number of measurements are made on individuals of a species, and a PCA is done. The first PC almost always has positive coefficients for all variables and simply reflects overall 'size' of the individuals. Later PCs usually contrast some of the measurements with others, and can often be interpreted as defining certain aspects of 'shape' which are important for the species. Blackith and Reyment (1971, Chapter 12) mention applications to squirrels, turtles, ammonites, foraminifera (marine micro-organisms) and various types of insects, but as a detailed example a small data set will be discussed, in which seven measurements were taken for a class of 28 students (15 women, 13 men). The seven measurements were circumferences of chest, waist, wrist and head, lengths of hand and forearm, and overall height. The PCA was done on the correlation matrix, even though it could be argued that, since all measurements were made in the same units, the covariance matrix might be more appropriate—see Sections 2.3 and 3.3. The correlation matrix was preferred because it was desired to treat all variables on an equal footing: the covariance matrix would have given greater weight to larger, and hence more variable, measurements, such as height and chest girth, and less weight to smaller measurements such as wrist girth and hand length.

Some of the results of the PCAs, done separately for women and men, are given in Tables 4.1 and 4.2. It can be seen that the form of the first two PCs is similar for the two sexes, with some similarity, too, for the third PC. Bearing in mind the small sample sizes, and the consequent large sampling variation in PC coefficients, it seems that the major sources of variation in the measurements, as given by the first three PCs, are similar for each sex. A

Table 4.1. First three PCs: student anatomical measurements.

Component number		1	2	3
			Women	
Hand		0.33	0.56	0.03
Wrist		0.26	0.62	0.11
Height		0.40	−0.44	−0.00
Forearm	Coefficients	0.41	−0.05	−0.55
Head		0.27	−0.19	0.80
Chest		0.45	−0.26	−0.12
Waist		0.47	0.03	−0.03
Eigenvalue		3.72	1.37	0.97
Cumulative percentage of total variation		53.2	72.7	86.5
			Men	
Hand		0.23	0.62	0.64
Wrist		0.29	0.53	−0.42
Height		0.43	−0.20	0.04
Forearm	Coefficients	0.33	−0.53	0.38
Head		0.41	−0.09	−0.51
Chest		0.44	0.08	−0.01
Waist		0.46	−0.07	0.09
Eigenvalue		4.17	1.26	0.66
Cumulative percentage of total variation		59.6	77.6	87.0

Table 4.2. Simplified version of the coefficients in Table 4.1.

Component number	1	2	3
		Women	
Hand	+	+	
Wrist	+	+	
Height	+	−	
Forearm	+		−
Head	+	(−)	+
Chest	+	(−)	
Waist	+		
		Men	
Hand	+	+	+
Wrist	+	+	−
Height	+	(−)	
Forearm	+	−	+
Head	+		−
Chest	+		
Waist	+		

combined PCA using all 28 observations would therefore seem appropriate, in order to get better estimates of the first three PCs. It is, of course, possible that later PCs are different for the two sexes, and that combining all 28 observations will obscure such differences. However, if we are interested solely in interpreting the first few, high variance, PCs, then this potential problem is relatively unimportant.

Before attempting to interpret the PCs, some explanation of Table 4.2 is necessary. Typically, computer packages which produce PCs give the coefficients to several decimal places. When interpreting PCs, as with other types of tabular data, it is only the general pattern of the coefficients which is really of interest, and not precise values to several decimal places. Table 4.1 gives only two decimal places and Table 4.2 simplifies still further. A '+' or '−' in Table 4.2 indicates a coefficient whose absolute value is greater than half the maximum coefficient (again in absolute value) for the relevant PC; the sign of the coefficient is also indicated. Similarly, a (+) or (−) indicates a coefficient whose absolute value is between a quarter and a half of the largest absolute value for the PC of interest. There are, of course, many ways of constructing a simplified version of the PC coefficients in Table 4.1. For example, another possibility is to rescale the coefficients in each PC so that the maximum value is ±1, and tabulate only the values of the coefficients, rounded to one decimal place, whose absolute values are above a certain cut-off, say 0.5 or 0.7. Values of coefficients below the cut-off are omitted, leaving blank spaces, as in Table 4.2. Some such simple representation is often helpful in interpreting PCs, particularly if a PCA is done on a large number of variables.

Sometimes a simplification such as that given in Table 4.2 may be rather too drastic, and it is therefore advisable to present coefficients rounded to one or two decimal places as well. Principal components with rounded coefficients will no longer be optimal, so that the variances of the first few will tend to be reduced, and exact orthogonality will be lost. However, it has been shown (Bibby, 1980; Green, 1977) that fairly drastic rounding of coefficients will make little difference to the variances of the PCs—see Section 12.4. Thus, presentation of rounded coefficients will still give linear functions of x with variances very nearly as large as those of the PCs, while at the same time easing interpretations.

Turning now to the interpretation of the PCs in the present example, the first PC clearly measures overall 'size' for both sexes, as would be expected (see Section 3.8) since all the correlations between the seven variables are positive, and it accounts for 53% (women) or 60% (men) of the total variation. The second PC for both sexes contrasts hand and wrist measurements with height, implying that, after overall size has been accounted for, the main source of variation is between individuals with large hand and wrist measurements relative to their heights, and individuals with the converse relationship. For women, head and chest measurements also have some contribution to this component, and for men the forearm measurement, which is closely related to height, partially replaces height in the component. This second PC accounts for slightly less than 20% of the total variation, for both sexes.

It should be noted that the sign of any PC is completely arbitrary. If every coefficient in a PC, $z_i = \mathbf{a}_i'\mathbf{x}$, has its sign reversed, the variance of z_i is unchanged, and so is the orthogonality of $\mathbf{a}_i$ with all other eigenvectors. For example, the second PC for men as recorded in Tables 4.1 and 4.2 has large positive values for students with large hand and wrist measurements relative to their height. If the sign of $\mathbf{a}_2$, and hence z_2, were reversed, the large positive values now occur for students with small hand and wrist measurements relative to height. The interpretation of the PC remains the same, even though the roles of 'large' and 'small' are reversed.

The third PCs differ more between the sexes but nevertheless retain some similarity. For women it is almost entirely a contrast between head and forearm measurements; for men these two measurements are also important, but, in addition, hand and wrist measurements appear with the same signs as forearm and head respectively. This component contributes 9%—14% of total variation.

Overall, the first three PCs account for a substantial proportion of total variation, 86.5% and 87.0% for women and men respectively. Although discussion of rules for deciding how many PCs to retain is deferred until Chapter 6, intuition strongly suggests that these percentages are large enough for three PCs to give an adequate representation of the data.

A similar, but much more substantial study, using seven measurements on 3000 criminals, was reported by Macdonell (1902) and is quoted by Maxwell (1977). The first PC again measured overall size, the second contrasted head and limb measurements, and the third could be readily interpreted as measuring the shape (roundness versus thinness) of the head. The percentages of total variation accounted for by each of the first three PCs were 54.3%, 21.4% and 9.3%, very similar to the proportions given in Table 4.1.

The sample size (28) is rather small in our example compared to that of Macdonnell's (1902), especially when the sexes are analysed separately, and caution would be needed in making any inference about the PCs in the population of students from which the sample is drawn. However, the same variables have been measured for other classes of students, and similar PCs have been found (see Sections 5.1 and 11.5). In any case, a description of the sample, rather than inference about the underlying population, is often what is required, and the PCs describe the major directions of variation within a sample, regardless of the sample size.

Various studies have been conducted on non-human species—see Blackith and Reyment (1971, Chapter 12) for references. Invariably the first PC measures overall size, and later PCs have often provided useful information regarding important sources of variation in shape, for various species. It should be noted that PCs provide only one way of looking at biological size and shape variation—see, for example, Sprent (1972) for a review of some other approaches to the subject, which is known as allometry.

4.2. The Elderly at Home

Hunt (1978) described a survey of the 'Elderly at Home' in which values of a large number of variables were collected for a sample of 2622 elderly individuals living in private households in the UK in 1976. The variables collected included standard demographic information of the type found in the decennial censuses, as well as information on dependency, social contact, mobility and income. As part of a project carried out for the Departments of the Environment and Health and Social Security, a PCA was done on a subset of 20 variables from Hunt's (1978) data. These variables are listed briefly in Table 4.3. Full details of the variables, and also of the project as a whole, are given by Jolliffe *et al.* (1982a), and shorter accounts of the main aspects of the project are available in Jolliffe *et al.* (1980, 1982b). It should be noted that many of the variables listed in Table 4.3 are discrete or even dichotomous. Some authors suggest that PCA should only be done on continuous variables, preferably with normal distributions. However, provided that inferential techniques which depend on assumptions such as multivariate normality (see Section 3.7) are not invoked, there is no real necessity for the variables to have any particular distribution. Admittedly, correlations or covariances, on which PCs are based, have particular relevance for normal random variables, but they are still valid for discrete variables provided that the possible values of the discrete variables have a genuine interpretation. Variables should not be defined with more than two possible values, unless the values have a valid meaning relative to each other. If 0, 1, 3 are possible values for a variable, then the values 1 and 3 must really be twice as far apart as the values 0 and 1. Further discussion of PCA, and related techniques, for discrete variables is given in Section 11.1.

It is widely accepted that old people who have only just passed retirement age, are different from the 'very old', so that it might be misleading to deal with all 2622 individuals together. Hunt (1978), too, recognized possible differences between age groups by taking a larger proportion of elderly whose

Table 4.3. Variables used in the PCA for the elderly at home.

1. Age	11. Separate kitchen
2. Sex	12. Hot water
3. Marital status	13. Car or van ownership
4. Employed	14. Number of elderly in household
5. Birthplace	15. Owner occupier
6. Father's birthplace	16. Council tenant
7. Length of residence in present household	17. Private tenant
8. Density: persons per room	18. Lives alone
9. Lavatory	19. Lives with spouse or sibling
10. Bathroom	20. Lives with younger generation

age was 75 or over in her sample, compared to those between 65 and 74, than is present in the population as a whole. It was therefore decided to analyse the two age groups 65–74 and 75+ separately, and part of each analysis consisted of a PCA on the correlation matrices for the 20 vairables listed in Table 4.3. It would certainly not be appropriate to use the covariance matrix when the variables are of several different types, as here.

It turned out that, for both age groups, as many as 11 PCs could be reasonably well interpreted, in the sense that not too many coefficients were far from zero. Because there are relatively few strong correlations between the 20 variables, the effective dimensionality of the 20 variables is around 10 or 11, a much less substantial reduction than in many other PCAs where there are large correlations between most of the variables (see Sections 4.3 and 6.4, for example). Eleven PCs accounted for 85.0% and 86.6% of the total variation for the 65–74 and 75+ age groups respectively.

Interpretations of the first 11 PCs for the two age groups are given in Table 4.4, together with the percentage of total variation given by each PC. The variances of corresponding PCs for the two age groups differ very little, and there are similar interpretations for several pairs of PCs, for example the first, second, sixth and eighth. In other cases there are groups of PCs involving the same variables, but in different combinations for the two age groups, for example the third, fourth and fifth PCs. Similarly, the ninth and tenth PCs involve the same variables for the two age groups, but the order of the PCs is reversed.

Principal component analysis has also been found useful in other demographic studies, one of the earliest being due to Moser and Scott (1961). In this study, there were 57 demographic variables measured for 157 British towns. A PCA of these data showed that, unlike the elderly data, dimensionality could be vastly reduced; there are 57 variables, but as few as four PCs account for 63% of the total variation. These PCs also have ready interpretations as measures of social class, population growth 1931–51, population growth after 1951, and overcrowding.

Similar studies have been done more recently on local authority areas in the UK, by Imber (1977) and Webber and Craig (1978)—see also Jolliffe *et al.* (1986). In each of these studies, as well as Moser and Scott (1961) and the 'elderly at home' project, the main objective was to classify the local authorities, towns or elderly individuals, and the PCA was done as a prelude to, or as part of cluster analysis. The use of PCA in cluster analysis is discussed further in Section 9.2, but the PCA in each study mentioned here provided useful information, separate from the results of the cluster analysis. For example, Webber and Craig (1978) used 40 variables, and they were able to interpret the first four PCs as measuring social dependence, family structure, age structure and industrial employment opportunity. These four components accounted for 29.5%, 22.7%, 12.0% and 7.4% respectively, so that 71.6% of the total variation is accounted for in four interpretable dimensions.

Table 4.4. Interpretations for the first 11 PCs for the 'elderly at home'.

65–74	75+
Component 1 (16.0%; 17.8%)*	
Contrasts single elderly living alone, with others.	Contrasts single elderly, particularly female, living alone, with others.
Component 2 (13.0%; 12.9%)	
Contrasts those lacking basic amenities (lavatory, bathroom, hot water) in private rented accommodation, with others.	Contrasts those lacking basic amenities (lavatory, bathroom, hot water), who also mainly lack a car and are in private, rented accommodation, not living with the next generation, with others.
Component 3 (9.5%; 10.1%)	
Contrasts council tenants, living in crowded conditions, with others.	Contrasts those who have a car, do not live in council housing (and tend to live in own accommodation) and tend to live with the next generation, with others.
Component 4 (9.2%; 9.2%)	
Contrasts immigrants living with next generation, with others. There are elements here of overcrowding and possession of a car.	Contrasts council tenants, mainly immigrant, living in crowded conditions, with others.
Component 5 (7.3%; 8.3%)	
Contrasts immigrants not living with next generation, with others. They tend to be older, fewer employed, fewer with a car, than in component 4.	Contrasts immigrants with others.
Component 6 (6.7%; 5.6%)	
Contrasts the younger employed people (tendency to be male), in fairly crowded conditions, often living with next generation, with others.	Contrasts younger (to a certain extent, male) employed with others.
Component 7 (5.6%; 5.1%)	
Contrasts long-stay people with a kitchen, with others.	Contrasts those lacking kitchen facilities with others. (NB: 1243 out of 1268 have kitchen facilities)
Component 8 (5.0%; 4.9%)	
Contrasts women living in private accommodation, with others.	Contrasts private tenants with others.
Component 9 (4.6%; 4.5%)	
Contrasts old with others.	Contrasts long-stay, mainly unemployed, individuals with others.

Table 4.4. (Continued).

65–74	75+
Component 10 (4.4%; 4.4%)	
Contrasts long-stay individuals, without a kitchen, with others.	Contrasts very old with others.
Component 11 (3.7%; 3.8%)	
Contrasts employed (mainly female) with others	Contrasts men with women.

* The two percentages are the percentages of variation accounted for by the relevant PC for the 65–74 and 75+ age groups respectively.

4.3. Spatial and Temporal Variation in Meteorology

Principal component analysis provides a widely-used method of describing patterns of pressure, temperature, or other meteorological variables over a large spatial area. For example, Richman (1983) states that, over the previous 3 years, more than 60 applications of PCA, or similar techniques, had appeared in meteorological/climatological journals. The example considered in detail in this section is taken from Maryon (1979) and is concerned with sea-level atmospheric pressure fields, averaged over half-month periods, for most of the Northern Hemisphere. There were 1440 half-months, corresponding to 60 years between 1900 and 1974, excluding the years 1916–21, 1940–48 when data were inadequate. The pressure fields are summarized by estimating average pressure at $p = 221$ grid points covering the Northern Hemisphere so that the data set consists of 1440 observations on 221 variables. Data sets of this size, or larger, are commonplace in meteorology, and a standard procedure is to replace the variables by a few large-variance PCs. The eigenvectors which define the PCs are often known as empirical orthogonal functions (EOFs) in the meteorological literature, and the values of the PCs are sometimes referred to as amplitude time series (Rasmusson et al., 1981) or, confusingly, as coefficients (Maryon, 1979).

For each PC, there is a coefficient (in the usual sense of the word) for each variable, and because variables are gridpoints (geographical locations) it is possible to plot each coefficient on a map at its corresponding gridpoint, and then draw contours through geographical locations having the same coefficient values. The map representation can greatly aid interpretation, as is illustrated in Figure 4.1. This figure, which comes from Maryon (1979), gives the map of coefficients (arbitrarily renormalized to give 'round numbers' on the contours) for the second PC from the pressure data set described above, and is much easier to interpret than would be the corresponding table of 221 coefficients. Half-months which have large positive scores for this PC will tend to have high values of the variables (i.e. high pressure values) where coefficients on the map are positive, and low values of the variables (low

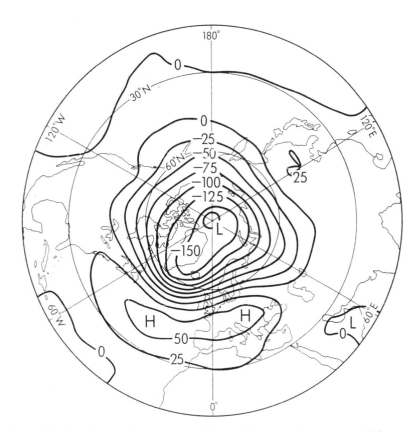

Figure 4.1. Graphical representation of the coefficients in the second PC for mean pressure data.

pressure values), at gridpoints where coefficients are negative. In Figure 4.1 this corresponds to low pressure in the polar regions, and high pressure in the subtropics, leading to situations where there is a strong westerly flow in high latitudes, at most longitudes. This is known as strong zonal flow, a reasonably frequent meteorological phenomenon, and the second PC therefore contrasts half-months with strong zonal flow with those of opposite character. Similarly, the first PC has one of its extremes identified as corresponding to an intense high pressure area over Asia and such situations are again a fairly frequent occurrence, although only in winter.

Several other PCs in Maryon's (1979) study can also be interpreted as corresponding to recognizable meteorological situations, especially when coefficients are plotted in map form. The use of PCs to summarize pressure fields, and other meteorological fields, has been found to be so valuable that it is now almost routine. For example, Craddock and Flood (1969) find PCs with ready interpretations for Northern Hemispheric 500 mb geopotential surfaces, Craddock and Flintoff (1970) do the same for 1000 mb surfaces and

1000–500 mb thickness, while more recently Overland and Preisendorfer (1982) interpret the first three PCs for data on spatial distributions of cyclone frequencies in the Bering Sea, whereas Wigley *et al.* (1984), among others, discuss PCs for European precipitation data, and Folland *et al.* (1985) find interpretable patterns in PCs of worldwide sea surface temperature anomalies.

Not only are the first few PCs readily interpreted in many meteorological examples, but they also frequently enable a considerable reduction to be made in the dimensions of the data set. In Maryon's (1979) study, for example, there are initially 221 variables, but 16 PCs account for over 90% of the total variation. Nor is this due to any disparity between variances causing a few dominant PCs; sizes of variance are fairly similar for all 221 variables.

Maryon's (1979) analysis was for a covariance matrix, which is reasonable since all variables are measured in the same units—see Sections 2.3 and 3.3. However, some climatologists have advocated using correlation, rather than covariance, matrices so that patterns of spatial correlation can be detected without possible domination by the stations and gridpoints with the largest variances—see Wigley *et al.* (1984).

4.4. Properties of Chemical Compounds

The main example given in this section is based on a subset of data given by Hansch *et al.* (1973), although the PCA was described by Morgan (1981). Seven properties (variables) were measured for each of 15 chemical substituents; the properties and substituents are listed in Table 4.5. Some of the results of a PCA for the correlation matrix for these data are given in Table 4.6.

Table 4.5. Variables and substituents considered by Hansch *et al.* (1973).

(a) Variables

1. π: Hansch's measure of lipophylicity

2. F $\rbrace$
3. R $\rbrace$ measures of electronic effect. F denotes 'field'; R denotes resonance

4. MR: molar refraction

5. σ_m $\rbrace$
6. σ_p $\rbrace$ further measures of electronic effect

7. MW: molecular weight

(b) Substituents

1. Br	2. Cl	3. F	4. I	5. CF_3
6. CH_3	7. C_2H_5	8. C_3H_7	9. C_4H_2	10. OH
11. NH_2	12. CH_2OH	13. SO_2CH_3	14. $SOCH_3$	15. $SO_2(NH_2)$

Table 4.6. First four PCs for Hansch *et al.*'s (1973) chemical data.

Component number		1	2	3	4
π		0.15	0.49	0.70	−0.45
F		−0.42	−0.36	0.34	0.13
R		−0.37	0.30	−0.44	−0.54
MR	Coefficients	−0.16	0.62	−0.23	0.49
σ_m		−0.48	−0.24	0.19	−0.03
σ_p		−0.50	0.01	−0.11	−0.30
MW		−0.40	0.30	0.31	0.40
Eigenvalue		3.79	1.73	0.74	0.59
Cumulative percentage of total variation		54.1	78.8	89.4	97.8

The aim of Hansch *et al.*'s work, and of much subsequent research in quantitative structure–activity relationships (QSAR), is to relate aspects of the structure of chemicals to their physical properties or activities, so that 'new' chemicals can be manufactured whose activities may be predicted in advance. Principal component analysis has played a rôle in recent papers on QSAR; for example, it has been used in conjunction with regression (see Chapter 8 and Mager (1980a)), and as a discriminant technique (see Section 9.1 and Mager (1980b)). Here we look only at the reduction of dimensionality and interpretations obtained by Morgan (1981) in this analysis of Hansch *et al.*'s data. The first two PCs in Table 4.6 account for 79% of the total variation, and the coefficients for each have a moderately simple structure. The first PC is essentially an average of all properties except π and MR, whereas the second PC is mainly an average of π and MR. Morgan (1981) also reports PCAs for a number of other similar data sets, in several of which the PCs have provided useful interpretations.

4.5. Stock Market Prices

The data in this example are the only set in this chapter which have previously appeared in a textbook (Press, 1972, Section 9.5.2). However, both the data and the PCs have interesting structures, and are worthy of further discussion. The data which were originally analysed by Feeney and Hester (1967) consist of 50 quarterly measurements between 1951 and 1963 of US stockmarket prices for the 30 industrial stocks which made up the Dow–Jones index at the end of 1961. Table 4.7 gives, in simplified form, the coefficients of the first two PCs, together with the percentage of variation accounted for by each PC, for both covariance and correlation matrices.

Looking first at the PCs for the correlation matrix, the first is a 'size'

Table 4.7. Coefficients for the first two PCs: stock market prices.

Component number	Correlation matrix		Covariance matrix	
	1	2	1	2
Allied Chemical	+	(−)		
Alcoa	+	−	(+)	+
American Can	+	−		
AT and T	+	+	(+)	−
American Tobacco	+	+		
Anaconda	(+)	−		+
Bethlehem	+	−		(+)
Chrysler	(−)	(−)		
du Pont	+	(−)	+	+
Eastman Kodak	+	(+)	+	−
General Electric	+		(+)	
General Foods	+	+	(+)	−
General Motors	+			
Goodyear	+			
International Harvester	+	(+)		
International Nickel	+		(+)	
International Paper	+	(−)		
Johns–Manville	+			
Owens–Illinois	+		(+)	
Proctor and Gamble	+	+	(+)	−
Sears	+	+	(+)	−
Standard Oil (Cal.)	+			
Esso	+	(−)		
Swift	(+)	(−)		
Texaco	+	(+)	(+)	(−)
Union Carbide	+	(−)	(+)	(+)
United Aircraft	+	−		+
US Steel	+	(−)	(+)	(+)
Westinghouse	+			
Woolworth	+	+		(−)
Percentage of variation accounted for	65.7	13.7	75.8	13.9

component, similar to those discussed in Section 4.1. It reflects the fact that all stock prices rose fairly steadily during the period 1951–63, with the exception of Chrysler. It accounts for roughly two-thirds of the variation in the 30 variables. The second PC can be interpreted as a contrast between 'consumer' and 'producer' stocks. 'Consumer' companies are those which mainly supply goods or services directly to the consumer, such as AT and T, American Tobacco, General Foods, Proctor and Gamble, Sears and Woolworth, whereas 'producer' companies sell their goods or services mainly to other companies, and include Alcoa, American Can, Anaconda, Bethlehem, Union Carbide and United Aircraft.

The PCs for the covariance matrix can be similarly interpreted, albeit with a change of sign for the second component, but the interpretation is slightly confounded, especially for the first PC, by the different-sized variances for each variable.

Feeney and Hester (1967) also performed a number of other PCAs using these and related data. In one analysis, they removed a linear trend from the stock prices before calculating PCs, and found that they had eliminated the size (trend) PC, and that the first PC was now very similar in form to the second PC in the original analyses. They also calculated PCs based on 'rate-of-return' rather than price, for each stock, and again found interpretable PCs. Finally, PCs were calculated for subperiods of 12 years of data, in order to investigate the stability of the PCs.

To conclude this example, note that it is of a special type, since each variable is a time series, in which consecutive observations are not independent. Further discussion of PCA for time series data is given in Section 11.2. A possible technique, for finding PCs which are free of the trend in a vector of time series, which is more general than that noted above for the present example, is described in Section 12.5.

CHAPTER 5

Graphical Representation of Data Using Principal Components

The main objective of a PCA is to reduce the dimensionality of a set of data. This is particularly advantageous if a set of data with many variables lies, in reality, close to a two-dimensional subspace (plane). In this case the data can be plotted with respect to these two dimensions, thus giving a straightforward visual representation of what the data look like, instead of having a large mass of numbers to digest. If the data fall close to a three-dimensional subspace it is still possible, with a little effort, to gain a good visual impression of the data, especially if a computer is available with interactive graphics. Even with slightly more dimensions it is possible, with some degree of ingenuity, to get a 'picture' of the data—see, for example, Chapters 10–12 (by Tukey and Tukey) in Barnett (1981)—although we shall concentrate almost entirely on two-dimensional representations in the present chapter.

If a good representation of the data exists in a small number of dimensions then PCA will find it, since the first q PCs give the 'best-fitting' q-dimensional subspace in the sense defined by Property G3 of Section 3.2. Thus, if we plot the values, for each observation, of the first two PCs, we get the best possible two-dimensional plot of the data; similarly for three or more dimensions. The first section of this chapter simply gives examples illustrating this procedure. We largely defer until the next chapter the problem of whether two PCs are adequate to represent most of the variation in the data, or whether we need more than two.

There are numerous other methods for representing high-dimensional data in two- or three-dimensions and, indeed, the book by Everitt (1978) is almost entirely on the subject, as are the conference proceedings edited by Wang (1978) and by Barnett (1981)—see also Chapter 5 of the book by Chambers *et al.* (1983). The only such techniques discussed in the present chapter will be those which have links with, or can be used in conjunction with, PCA.

Section 5.2 discusses principal co-ordinate analysis, which constructs low-dimensional plots of a set of data from information about similarities or dissimilarities between pairs of observations. It turns out that the plots given by this analysis are equivalent to plots with respect to PCs in certain special cases.

The biplot, described in Section 5.3, is also closely related to PCA. There are a number of variants of the biplot idea, but all give a simultaneous display of n observations and p variables on the same two-dimensional diagram. In one of the variants, the plot of observations is identical to a plot with respect to the first two PCs, but the biplot simultaneously gives graphical information about the relationships between variables. The relative positions of variables *and* observations, which are plotted on the same diagram, can also be interpreted.

Correspondence analysis, which is discussed in Section 5.4, again gives two-dimensional plots, but only for data of a special form. Whereas PCA and the biplot operate on a matrix of n observations on p variables, and principal co-ordinate analysis and other types of scaling or ordination techniques use data in the form of a similarity or dissimilarity matrix, correspondence analysis is used on contingency tables, i.e. data classified according to two categorical variables. The link with PCA is less straightforward than for principal co-ordinate analysis or the biplot, but the ideas of PCA and correspondence analysis have some definite connections. In Section 5.5 some comparisons are made, briefly, between PCA and the other techniques introduced in this chapter.

The final section of this chapter discusses some methods which have been used for representing multivariate data in two dimensions when more than two or three PCs are needed to give an adequate representation of the data. The first q PCs can still be helpful in reducing the dimensionality in such cases, even when q is much larger than 2 or 3.

Finally, note that as well as the graphical representations described in the present chapter, we have already seen, in Section 4.3, one other type of plot which uses PCs. This type of plot is rather specialized, but is used extensively in meteorology and climatology.

5.1. Plotting Data with Respect to the First Two (or Three) Principal Components

The idea here is simple: if a data set $\{x_1, x_2, \ldots, x_n\}$ has p variables, then the observations can be plotted as points in p-dimensional space. If we wish to plot the data in a 'best-fitting' q-dimensional subspace $(q < p)$, where 'best-fitting' is defined, as in Property G3 of Section 3.2, as minimizing the sum of squared perpendicular distances of $x_1, x_2, \ldots, x_n$ from the subspace, then the appropriate subspace is defined by the first q PCs.

Two-dimensional plots are particularly useful for detecting patterns in the data, and three-dimensional plots or models, though generally less easy to interpret quickly, can sometimes give additional insights. If the data do not lie close to a two- (or three-) dimensional subspace, then no two- (or three-) dimensional plot of the data will provide an adequate representation (although Section 5.6 discusses briefly the use of indirect ways for presenting the data in two dimensions in such cases). However, if the data are close to a q-dimensional subspace, then most of the variation in the data will be accounted for by the first q PCs, and a plot of the observations with respect to these PCs will give a realistic picture of what the data look like, unless important aspects of the data structure are concentrated in the direction of low variance PCs. Plotting data sets with respect to the first two PCs is now illustrated by two examples, with further illustrations given, in conjunction with other examples, later in this chapter, and in subsequent chapters.

5.1.1. Examples

Two examples are given here in order to illustrate the sort of interpretation which may be given to plots of observations with respect to their first two PCs. These two examples do not reveal any strong, but previously unknown, structure such as clusters; examples illustrating clusters will be presented in Section 9.2. Nevertheless, useful information can still be gleaned from the plots.

Anatomical Measurements

The data presented here consist of the same seven anatomical measurements as in the data set of Section 4.1, but for a different set of students, this time comprising 11 women and 17 men. A PCA was done on the correlation matrix for all 28 observations and, as in the analyses of Section 4.1 for each sex separately, the first PC is an overall measurement of size. The second PC is a contrast between the head measurement and the other six variables, and is therefore not particularly similar to any of the first three PCs for the separate sexes found in Section 4.1, though it is closest to the third component for the women. The difference between the second PC and those from the earlier analyses may be partially due to the fact that the sexes have been combined, but it is also likely to reflect some instability in all but the first PC due to relatively small sample sizes. The first two PCs for the present data account for 69% and 11% of the total variation respectively, so that a two-dimensional plot with respect to these PCs, representing 80% of the variation, will give a reasonably good approximation to the relative positions of the observations in seven-dimensional space.

Figure 5.1 gives three plots of the observations with respect to the first

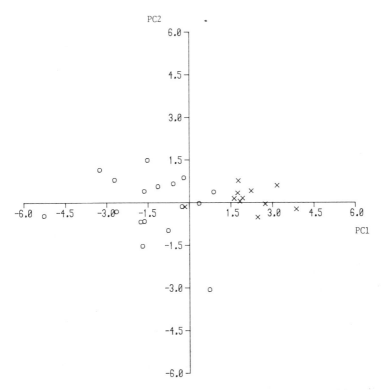

Figure 5.1(a). Student anatomical measurements: plot of 28 students with respect to their first two PCs. × denotes women; o denotes men.

two PCs. In Figure 5.1(a) it can be seen that there is one clear outlier with respect to the second PC, at the bottom of the plot. A second observation, at the left of the plot, is rather extreme on the first PC. These two observations and other potential outliers will be discussed further in Section 10.1. The observation at the bottom of the diagram has such an extreme value for the second PC (roughly twice as large, in absolute terms, as any other observation) that it could be mainly responsible for the second PC taking the form which it does. This possibility will be discussed further in Section 10.2.

Figure 5.1(b), (c) is the same as Figure 5.1(a), except that superimposed on it are convex hulls for the two groups, men and women (Figure 5.1(b)), and the minimum spanning tree (Figure 5.1(c)). Convex hulls are useful in indicating the areas of a two-dimensional plot covered by various subsets of observations. Here they demonstrate that, although the areas covered by men and women overlap slightly, the two sexes largely occupy different areas of the diagrams. The separation is mainly in terms of the first PC (overall size) with very little differentiation between sexes on the second PC. The plot therefore displays the (unsurprising) result that the two groups of observations corresponding to the two sexes differ mainly in terms of overall size.

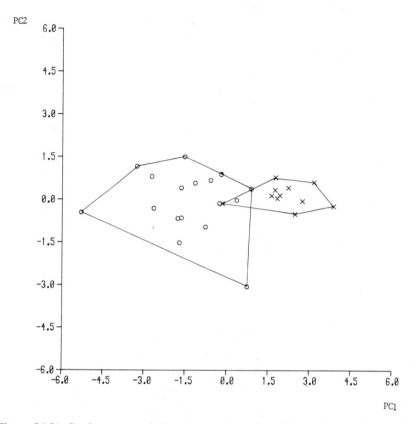

Figure 5.1(b). Student anatomical measurements: plot of 28 students with convex hulls for men and women superimposed.

It was noted above that the two-dimensional plot represents 80% of the total variation of the 28 observations in seven-dimensional space. Percentage of total variation is an obvious measure of how good is the two-dimensional representation, but many of the other criteria which are discussed in Section 6.1 could be used instead. Alternatively, an informal way of judging the goodness-of-fit in two dimensions is to superimpose a minimum spanning tree (MST) on the diagram, as in Figure 5.1(c). The MST is a set of lines drawn between pairs of points such that

(i) each point is connected to every other point by a sequence of lines;
(ii) there are no closed loops;
(iii) the sum of 'lengths' of lines is minimized.

If the 'lengths' of the lines are defined as distances in seven-dimensional space, then the corresponding MST will give an indication of the closeness-of-fit of the two-dimensional representation. For example, it is seen that

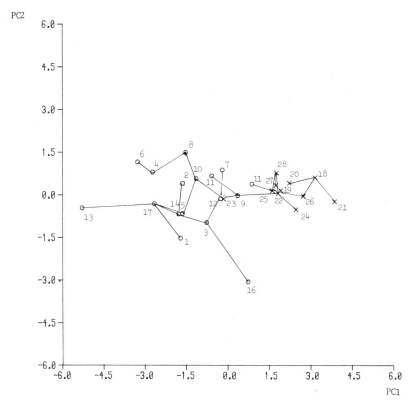

Figure 5.1(c). Student anatomical measurements: plot of 28 students with minimum spanning tree superimposed.

observations 5 and 14, which are very close in two dimensions, are joined via observation 17, and so must both be closer to observation 17 in seven-dimensional space than to each other. There is therefore some distortion in the two-dimensional representation in the vicinity of observations 5 and 14. Similar remarks apply to observations 12 and 23, and to the group of observations 19, 22, 25, 27, 28. However, there appears to be little distortion for the better-separated observations.

Artistic Qualities of Painters

The second data set described in this section was discussed by Davenport and Studdert-Kennedy (1972). It consists of a set of subjective measurements of the artistic qualities, 'composition', 'drawing', 'colour' and 'expression' for 54 painters. The measurements, on a scale from 0 to 20, were compiled in France in 1708 by Roger de Piles for painters 'of established reputation'.

Table 5.1. First two PCs: artistic qualities of painters.

		Component 1	Component 2
Composition		0.50	−0.49
Drawing	Coefficients	0.56	0.27
Colour		−0.35	−0.77
Expression		0.56	−0.31
Eigenvalue		2.27	1.04
Cumulative percentage of total variation		56.8	82.8

Davenport and Studdert-Kennedy (1972) give data for 56 painters but, like them, the present analysis omits two painters for whom one measurement is missing.

Table 5.1 gives the variances and coefficients for the first two PCs, based on the correlation matrix, for the 54 painters with complete data. The components, and their contributions to the total variation, are very similar to those found by Davenport and Studdert-Kennedy (1972) for the covariance matrix. This strong similarity between the PCs for correlation and covariance matrices is relatively unusual (see Section 3.3) and is due to the near-equality of the variances for the four variables. The first component is interpreted by Davenport and Studdert-Kennedy (1972) as an index of de Piles' overall assessment of the painters, although the negative coefficient for colour needs some additional explanation. The form of this first PC could be predicted from the correlation matrix. If the sign of the variable 'colour' is changed, then all correlations in the matrix are positive, so that we would expect the first PC to have positive coefficients for all variables, after this redefinition of 'colour'—see Section 3.8. The second PC has its largest coefficient for colour, but the other coefficients are also non-negligible.

A plot of the 54 painters with respect to the first two components is given in Figure 5.2, and this two-dimensional display describes 82.8% of the total variation. The main feature of Figure 5.2 is that painters of the same 'school' are mostly fairly close to each other. For example, the ten 'Venetians' {Bassano, Bellini, Veronese, Giorgione, Murillo, Palma Vecchio, Palma Giovane, Pordenone, Tintoretto, Titian}, are indicated on the figure, and are all in a relatively small area in the bottom left of the plot. Davenport and Studdert-Kennedy (1972) perform a cluster analysis on the data, and display the clusters on a plot of the first two PCs. The clusters dissect the data in a sensible looking manner, and none of them has a convoluted shape on the PC plot. However, there is little evidence of a strong cluster structure on Figure 5.2. Possible exceptions are a group of three isolated painters near the bottom of the plot, and four painters at the extreme left. The first group are all members of the 'Seventeenth Century School', namely Rembrandt, Rubens, and Van Dyck, and the second group consists of three 'Venetians', Bassano, Bellini, Palma Vecchio, together with the 'Lombardian' Caravaggio.

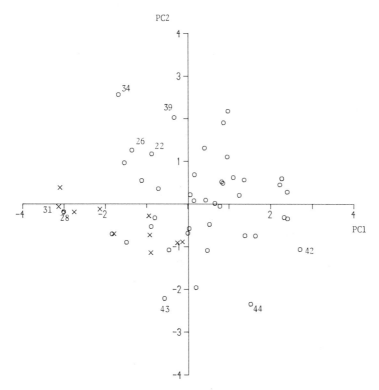

Figure 5.2. Artistic qualities of painters: plot of 54 painters with respect to their first two PCs. × denotes member of the 'Venetian' school.

This data set will be discussed again in Sections 5.3 and 10.2, and the numbered observations on Figure 5.2 will be referred to there.

Further examples of the use of PCA in conjunction with cluster analysis are given in Section 9.2.

5.2. Principal Co-ordinate Analysis

Principal co-ordinate analysis is a scaling or ordination method and was popularized by Gower (1966). Torgerson (1958) discussed similar ideas, but did not point out the links between principal co-ordinate analysis and PCA which were noted by Gower. Like the more widely-known non-metric multi-dimensional scaling (Kruskal, 1964a, b), the technique starts with a matrix of similarities or dissimilarities between a set of observations and aims to produce a low-dimensional graphical plot of the data in such a way that distances between points in the plot are close to original dissimilarities. Thus the starting point (a $(n \times n)$ matrix of (dis)similarities) for principal co-ordinate analysis is different from that of PCA, which starts with the $(n \times p)$ data matrix. However, in special cases discussed below the two techniques give precisely the same low-dimensional representation. Before showing the equivalence between PCA and principal co-ordinate analysis in these special cases we need first to describe principal co-ordinate analysis in some detail.

Suppose that $\mathbf{T}$ is a $(n \times n)$ positive-semidefinite symmetric matrix of similarities between a set of n observations. (Note that it is fairly standard notation to use $\mathbf{A}$, rather than $\mathbf{T}$, here. However, we have avoided the use of $\mathbf{A}$ in this context, since it is consistently taken to be the matrix of PC coefficients throughout the current text.) From the spectral decomposition of $\mathbf{T}$ (Property A3 of Sections 2.1 and 3.1 gives the spectral decomposition of a covariance matrix, but the same idea is valid for any symmetric matrix), we have

$$\mathbf{T} = \tau_1 \mathbf{b}_1 \mathbf{b}_1' + \tau_2 \mathbf{b}_2 \mathbf{b}_2' + \cdots + \tau_n \mathbf{b}_n \mathbf{b}_n', \tag{5.2.1}$$

where $\tau_1 \geq \tau_2 \geq \cdots \geq \tau_n$ are the eigenvalues of $\mathbf{T}$ and $\mathbf{b}_1, \mathbf{b}_2, \ldots, \mathbf{b}_n$ are the corresponding eigenvectors. Alternatively, this may be written

$$\mathbf{T} = \mathbf{c}_1 \mathbf{c}_1' + \mathbf{c}_2 \mathbf{c}_2' + \cdots + \mathbf{c}_n \mathbf{c}_n', \tag{5.2.2}$$

where

$$\mathbf{c}_j = \tau_j^{1/2} \mathbf{b}_j, \qquad j = 1, 2, \ldots, n.$$

Now consider the n observations as points in n-dimensional space with the jth co-ordinate for the ith observation equal to c_{ij}, the ith element of $\mathbf{c}_j$. With this geometric interpretation of the n observations, the Euclidean distance between the hth and ith observations is

$$\Delta_{hi}^2 = \sum_{j=1}^{n} (c_{hj} - c_{ij})^2$$

$$= \sum_{j=1}^{n} c_{hj}^2 + \sum_{j=1}^{n} c_{ij}^2 - 2 \sum_{j=1}^{n} c_{hj} c_{ij}.$$

But from (5.2.2), the (h, i)th element of $\mathbf{T}$ can be written

$$t_{hi} = \sum_{j=1}^{n} c_{hj} c_{ij}, \qquad h, i = 1, 2, \ldots, n,$$

so

$$\Delta_{hi}^2 = t_{hh} + t_{ii} - 2t_{hi}.$$

Principal co-ordinate analysis then attempts to find the 'best-fitting' q-dimensional $(q < n)$ approximation to the n-dimensional representation defined above. 'Best-fitting' is defined here in the same way as in the geometric definition of PCA (Property G3 of Section 3.2), so that 'PCs' are now found for the n 'observations' defined in n dimensions by the co-ordinates c_{ij}. A q-dimensional principal co-ordinate representation is then given by plotting the co-ordinates of the observations with respect to the first q PCs. Principal co-ordinate analysis therefore consists of two stages, both of which involve finding eigenvalues and eigenvectors of $(n \times n)$ matrices:

(i) find the eigenvectors $(c_1, c_2, \ldots, c_n)$ of T and represent the n observations as points in n-dimensional space with co-ordinate c_{ij} for the ith observation in the jth dimension;

(ii) find the PCs for the 'data set' in n dimensions defined in (i), and calculate co-ordinates of the n observations with respect to the first q PCs.

If the vectors c_j defined in the first stage have $\sum_{i=1}^{n} c_{ij} = 0$, then the covariance matrix which is calculated in (ii) will be proportional to $C'C$ where C is the $(n \times n)$ matrix with jth column, $c_j, j = 1, 2, \ldots, n$. But

$$c_j'c_k = \begin{cases} \tau_j, & j = k, \\ 0, & j \neq k, \end{cases}$$

since the eigenvectors in the spectral decomposition (5.2.1) have the property

$$b_j'b_k = \begin{cases} 1, & j = k, \\ 0, & j \neq k, \end{cases}$$

and

$$c_j = \tau_j^{1/2} b_j, \qquad j = 1, 2, \ldots, n.$$

$C'C$ will therefore be a diagonal matrix with diagonal elements $\tau_j, j = 1, 2, \ldots$, n, so that the first q principal co-ordinates of the n observations are simply the values of c_{ij} for $i = 1, 2, \ldots, n; j = 1, 2, \ldots, q$. Thus when $\sum_{i=1}^{n} c_{ij} = 0$, the second stage (ii) is unnecessary.

In general, although a similarity matrix T need not lead to $\sum_{i=1}^{n} c_{ij} = 0$, this property can be readily achieved by replacing T by an adjusted similarity matrix. In this adjustment, t_{hi} is replaced by $t_{hi} - \bar{t}_h - \bar{t}_i - \bar{t}$, where $\bar{t}_i$ denotes the mean of the elements in the hth row (or column, since T is symmetric) of T, and $\bar{t}$ is the mean of all elements in T. This adjusted similarity matrix has $\sum_{i=1}^{n} c_{ij} = 0$, and gives the same value of Δ_{hi}^2 for each pair of observations as does T (Gower, 1966). Thus we can replace the second stage of principal co-ordinate analysis by an initial adjustment of T, for any similarity matrix T.

Principal co-ordinate analysis is equivalent to a plot with respect to the first q PCs, when the measure of similarity between two points is proportional to $-d_{hi}^2$, where d_{hi}^2 is the Euclidean squared distance between the hth and ith observations, calculated from the usual $(n \times p)$ data matrix. Assume $t_{hi} = -\gamma d_{hi}^2$, where γ is a positive constant; then if stage (i) of a principal co-ordinate analysis is carried out, the 'distance' between a pair of points in the constructed n-dimensional space is

$$\Delta_{hi}^2 = (t_{hh} + t_{ii} - 2t_{hi})$$

$$= \gamma(-d_{hh}^2 - d_{ii}^2 + 2d_{hi}^2)$$

$$= 2\gamma d_{hi}^2,$$

since Euclidean distance from a point to itself is zero. Thus, apart from a possible rescaling if γ is taken to be a value other than $\frac{1}{2}$, the first stage of principal co-ordinate analysis has correctly reproduced the relative positions of the n observations (which will in fact lie in a p-dimensional subspace of n-dimensional space), so that the subsequent PCA in stage (ii) will give the same result as a PCA on the original data.

Two special cases of this result are of interest; first, consider the situation where all variables are binary. A commonly used measure of similarity between individuals h and i is the proportion of the p variables for which h and i take the same value, and it can be easily demonstrated (Gower, 1966), that this measure is equivalent to Euclidean distance. Thus, although PCA of discrete, and in particular binary, data has its critics, it is equivalent to principal co-ordinate analysis with a very plausible measure of similarity. Principal component analysis for discrete data is discussed further in Section 11.1.

The second special case occurs when the elements of the similarity matrix $\mathbf{T}$ are defined as 'covariances' between observations, so that $\mathbf{T}$ is proportional to $\mathbf{XX'}$, where $\mathbf{X}$, as before, is the $(n \times p)$ matrix whose (i, j)th element is the value of the jth variable, measured about its mean $\bar{x}_j$, for the ith observation. In this case the (h, i)th similarity is (apart from a constant)

$$t_{hi} = \sum_{j=1}^{p} x_{hj}x_{ij}$$

and the distance between the points in the n-dimensional space constructed in the first stage of the principal co-ordinate analysis is

$$\Delta_{hi}^2 = t_{hh} + t_{ii} - 2t_{hi}$$

$$= \sum_{j=1}^{p} x_{hj}^2 + \sum_{j=1}^{p} x_{ij}^2 - 2\sum_{j=1}^{p} x_{hj}x_{ij}$$

$$= \sum_{j=1}^{p} (x_{hj} - x_{ij})^2$$

$$= d_{hi}^2,$$

the Euclidean distance between the observations using the original p variables. As before, the PCA in the second stage of principal co-ordinate analysis will give the same results as a PCA on the original data.

Even in cases where PCA and principal co-ordinate analysis give equivalent two-dimensional plots there is a difference, namely that in principal co-ordinate analysis there are no vectors of coefficients defining the axes in terms of the original variables. This means that, unlike PCA, the axes cannot be interpreted, unless the corresponding PCA is also done.

The equivalence between PCA and principal co-ordinate analysis in the circumstances described above is termed a *duality* between the two techniques by Gower (1966). The techniques are dual in the sense that PCA operates on a matrix of similarities between variables, whereas principal co-ordinate analysis operates on a matrix of similarities between observations (individuals), but both can lead to equivalent results.

To summarize, principal co-ordinate analysis gives a low-dimensional representation of data, when the data are given in the form of a similarity or dissimilarity matrix. Since it can be used with any form of similarity or dissimilarity matrix, it is, in one sense, 'more powerful' than, and 'extends', PCA (Gower, 1967). However, as will be seen in subsequent chapters, PCA has many uses other than representing data graphically, which is the over-riding purpose of principal co-ordinate analysis.

Except in the special cases discussed above, principal co-ordinate analysis has no direct relationship with PCA, so no examples will be given of the general application of the technique. In the case where principal co-ordinate analysis gives an equivalent representation to that of PCA, nothing new would be demonstrated by giving additional examples. The examples given in Section 5.1 (and elsewhere) which are presented as plots with respect to the first two PCs are, in fact, equivalent to two-dimensional principal co-ordinate plots if the 'dissimilarity' between observations h and i is proportional to d_{hi}^2, the Euclidean squared distance between the hth and ith observations in p dimensions.

In most cases, if the data are available in the form of an $(n \times p)$ matrix of p variables measured for each of n observations there is no advantage in doing a principal co-ordinate analysis instead of a PCA. However, an exception occurs when $n < p$, especially if $n \ll p$ as happens for some types of chemical, meteorological and biological data. Since principal co-ordinate analysis and PCA find eigenvectors of a $(n \times n)$ matrix and a $(p \times p)$ matrix respectively, the dual analysis based on principal co-ordinates will have computational advantages in such cases.

5.3. Biplots

The previous two sections describe plots of the n *observations*, usually in two dimensions. Biplots similarly provide plots of the n observations, but simultaneously they give plots of the relative positions of the p *variables* in two dimensions. Furthermore, if the two types of plot are superimposed, this will provide additional information about relationships between variables and observations, which is not available in either of the individual plots. Correspondence analysis, which is discussed in the next section, also provides two superimposed plots, but it is appropriate for a different type of data structure.

Biplots have been largely developed by Gabriel (1971, and several subsequent papers) and are based on the singular value decomposition (SVD), which was described in Section 3.5. This states that the $(n \times p)$ matrix $\mathbf{X}$ of n

observations on p variables, measured about their sample means, can be written

$$X = ULA', \tag{5.3.1}$$

where U, A are $(n \times r)$, $(p \times r)$ matrices respectively, each with orthonormal columns, L is a $(r \times r)$ diagonal matrix with elements $l_1^{1/2} \geq l_2^{1/2} \geq \cdots \geq l_r^{1/2}$, and r is the rank of X. Now define L^α, for $0 \leq \alpha \leq 1$, as the diagonal matrix whose elements are $l_1^{\alpha/2}, l_2^{\alpha/2}, \ldots, l_r^{\alpha/2}$, with a similar definition for $L^{1-\alpha}$, and let $G = UL^\alpha$, $H' = L^{1-\alpha}A'$. Then

$$GH' = UL^\alpha L^{1-\alpha}A'$$

$$= ULA' = X,$$

and the (i, j)th element of X can be written

$$x_{ij} = g_i' h_j, \tag{5.3.2}$$

where g_i', $i = 1, 2, \ldots, n$, and h_j', $j = 1, 2, \ldots, p$, are the rows of G and H respectively. Both the g_i's and h_j's have r elements, and if X has rank 2, all could be plotted as points in two-dimensional space. In the more general case where $r > 2$, it was noted in Section 3.5 that (5.3.1) can be written

$$x_{ij} = \sum_{k=1}^{r} u_{ik} l_k^{1/2} a_{jk}, \tag{5.3.3}$$

which can often be well approximated by

$$_m\tilde{x}_{ij} = \sum_{k=1}^{m} u_{ik} l_k^{1/2} a_{jk}, \quad \text{with} \quad m < r. \tag{5.3.4}$$

But (5.3.4) can be written

$$_m\tilde{x}_{ij} = \sum_{k=1}^{m} g_{ik} h_{jk}$$

$$= g_i^{*\prime} h_j^*,$$

where g_i^*, h_j^* contain the first m elements of g_i and h_j respectively. In the case where (5.3.4) with $m = 2$ provides a good approximation to (5.3.3), g_i^*, $i = 1, 2, \ldots, n$, h_j^*, $j = 1, 2, \ldots, p$ will together give a good two-dimensional representation of both the n observations and the p variables. This type of approximation could, of course, be used for values of $m > 2$, but the graphical representation would then be less clear. Gabriel (1981) refers to the extension to $m \geq 3$ as a *bimodel*, reserving the term 'biplot' for the case where $m = 2$.

In the description of biplots above there is an element of non-uniqueness since the scalar α which occurs in the definition of G and H can take any value between 0 and 1 (or presumably outside that range) and still lead to a factorization of the form (5.3.2). Two particular values of α, $\alpha = 0$ and $\alpha = 1$, lead to especially useful interpretations for the biplot.

If $\alpha = 0$, then $\mathbf{G} = \mathbf{U}$ and $\mathbf{H}' = \mathbf{LA}'$ or $\mathbf{H} = \mathbf{AL}$. This means that

$$\mathbf{X}'\mathbf{X} = (\mathbf{GH}')'(\mathbf{GH}')$$

$$= \mathbf{HG}'\mathbf{GH}'$$

$$= \mathbf{HU}'\mathbf{UH}'$$

$$= \mathbf{HH}',$$

because the columns of $\mathbf{U}$ are orthonormal. The product $\mathbf{h}_j'\mathbf{h}_k$ is therefore equal to $(n-1)$ times the covariance, s_{jk}, between the jth and kth variables, and $\mathbf{h}_j^{*\prime}\mathbf{h}_k^*$, where $\mathbf{h}_j^*$, $j = 1, 2, \ldots, p$ are as defined below (5.3.4), provides an approximation to $(n-1)s_{jk}$. The lengths $\mathbf{h}_j'\mathbf{h}_j$ of the vectors $\mathbf{h}_j$, $i = 1, 2, \ldots, p$ will represent the variances of the variables, $x_1, x_2, \ldots, x_p$, and the cosines of the angles between the $\mathbf{h}_j$'s will represent correlations between variables. Plots of the $\mathbf{h}_j^*$'s will therefore provide a two-dimensional picture (usually an approximation, but often a good one) of the elements of the covariance matrix $\mathbf{S}$, and such plots are advocated by Corsten and Gabriel (1976) as a means of comparing the variance–covariance structures of several different data sets. An earlier paper, by Gittins (1969), which is reproduced in Bryant and Atchley (1975), also gives plots of the $\mathbf{h}_j^*$'s, although it does not discuss their formal properties.

Not only do the $\mathbf{h}_j$'s have a ready graphical interpretation when $\alpha = 0$, but the $\mathbf{g}_i$'s also have the satisfying property that the Euclidean distance between $\mathbf{g}_h$ and $\mathbf{g}_i$ in the biplot is proportional to the Mahalanobis distance between the hth and ith observations in the complete data set. The Mahalanobis distance between two observations $\mathbf{x}_h$, $\mathbf{x}_i$, assuming that $\mathbf{X}$ has rank p, so that $\mathbf{S}^{-1}$ exists, is defined as

$$\delta_{hi}^2 = (\mathbf{x}_h - \mathbf{x}_i)'\mathbf{S}^{-1}(\mathbf{x}_h - \mathbf{x}_i), \tag{5.3.5}$$

and is often used as an alternative to the Euclidean distance $d_{hi}^2 = (\mathbf{x}_h - \mathbf{x}_i)'(\mathbf{x}_h - \mathbf{x}_i)$. Whereas Euclidean distance treats all variables on an equal footing, which essentially assumes that all variables have equal variances and are uncorrelated, Mahalanobis distances gives relatively less weight to variables with large variances, and to groups of highly correlated variables.

Rewriting (5.3.2) as $\mathbf{x}_i' = \mathbf{g}_i'\mathbf{H}'$, $i = 1, 2, \ldots, n$, and substituting in (5.3.5) gives

$$\delta_{hi}^2 = (\mathbf{g}_h - \mathbf{g}_i)'\mathbf{H}'\mathbf{S}^{-1}\mathbf{H}(\mathbf{g}_h - \mathbf{g}_i)$$

$$= (n-1)(\mathbf{g}_h - \mathbf{g}_i)'\mathbf{LA}'(\mathbf{X}'\mathbf{X})^{-1}\mathbf{AL}(\mathbf{g}_h - \mathbf{g}_i), \tag{5.3.6}$$

since $\mathbf{H}' = \mathbf{LA}'$ and $\mathbf{S}^{-1} = (n-1)(\mathbf{X}'\mathbf{X})^{-1}$.
But

$$\mathbf{X}'\mathbf{X} = (\mathbf{ULA}')'(\mathbf{ULA}')$$

$$= \mathbf{AL}(\mathbf{U}'\mathbf{U})\mathbf{LA}'$$

$$= \mathbf{AL}^2\mathbf{A}',$$

and

$$(\mathbf{X}'\mathbf{X})^{-1} = \mathbf{AL}^{-2}\mathbf{A}'.$$

Substituting in (5.3.6) gives

$$\delta_{hi}^2 = (n-1)(\mathbf{g}_h - \mathbf{g}_i)'\mathbf{L}(\mathbf{A}'\mathbf{A})\mathbf{L}^{-2}(\mathbf{A}'\mathbf{A})\mathbf{L}(\mathbf{g}_h - \mathbf{g}_i)$$

$$= (n-1)(\mathbf{g}_h - \mathbf{g}_i)'\mathbf{LL}^{-2}\mathbf{L}(\mathbf{g}_h - \mathbf{g}_i), \quad \text{since the columns of } \mathbf{A} \text{ are orthonormal,}$$

$$= (n-1)(\mathbf{g}_h - \mathbf{g}_i)'(\mathbf{g}_h - \mathbf{g}_i), \quad \text{as required.}$$

An adaptation to the straightforward factorization given above for $\alpha = 0$ will improve the interpretation of the plot still further. If we multiply the $\mathbf{g}_i$'s by $(n-1)^{1/2}$ and, correspondingly, divide the $\mathbf{h}_j$'s by $(n-1)^{1/2}$, then the distances between the modified $\mathbf{g}_i$'s will be *equal* (not just proportional) to the Mahalanobis distance and, if $m = 2 < p$, then the Euclidean distance between $\mathbf{g}_h^*$ and $\mathbf{g}_i^*$ will give an easily visualized approximation to the Mahalanobis distance between $\mathbf{x}_h$ and $\mathbf{x}_i$. Furthermore, the lengths $\mathbf{h}_j'\mathbf{h}_j$ will be *equal* to variances of the variables. This adaptation was noted by Gabriel (1971), and is used in the examples below.

A further interesting property of the biplot when $\alpha = 0$ is that measures can be written down of how well the plot approximates

(a) the centred data matrix $\mathbf{X}$;
(b) the covariance matrix $\mathbf{S}$; and
(c) the matrix of Malalanobis distances between each pair of observations.

These measures are, respectively (Gabriel 1971)

(a) $(l_1 + l_2) \Big/ \sum_{k=1}^{r} l_k$;

(b) $(l_1^2 + l_2^2) \Big/ \sum_{k=1}^{r} l_k^2$;

(c) $(l_1^0 + l_2^0) \Big/ \sum_{k=1}^{r} l_k^0 = 2/r.$

Because $l_1 \geq l_2 \geq \cdots \geq l_r$, these measures imply that the biplot gives a better approximation to the variances and covariances than to the (Mahalanobis) distances between observations. This is in contrast to principal co-ordinate plots which concentrate on giving as good a fit as possible to interobservation dissimilarities or distances, and do not consider directly the elements of $\mathbf{X}$ or $\mathbf{S}$.

We have now seen readily interpretable properties of both the $\mathbf{g}_i^*$'s and the $\mathbf{h}_j^*$'s separately for the biplot when $\alpha = 0$, but there is a further property, valid for any value of α, which shows that the plots of the $\mathbf{g}_i^*$'s and $\mathbf{h}_j^*$'s can be usefully superimposed, rather than simply considering them separately.

Because of the relationship $x_{ij} = \mathbf{g}'_i \mathbf{h}_j$, x_{ij} is represented by the projection of $\mathbf{g}_i$ onto $\mathbf{h}_j$. Remembering that x_{ij} is the value, for the ith observation, of the jth variable measured about its sample mean, values of x_{ij} close to zero (i.e. observations close to the sample mean of the jth variable) will only be achieved if $\mathbf{g}_i$ and $\mathbf{h}_j$ are nearly orthogonal. Conversely, observations for which x_{ij} is a long way from zero will have $\mathbf{g}_i$ lying in a similar direction to $\mathbf{h}_j$. The relative positions of the points defined by the $\mathbf{g}_i$'s and $\mathbf{h}_j$'s, or their approximations in two dimensions, the $\mathbf{g}_i^*$'s and $\mathbf{h}_j^*$'s, will therefore give information about which observations take large, average and small values on each variable.

Turning to the biplot with $\alpha = 1$, the properties relating to $\mathbf{g}_i$ and $\mathbf{h}_j$ separately will change. With $\alpha = 1$, we have

$$\mathbf{G} = \mathbf{UL}, \qquad \mathbf{H}' = \mathbf{A}',$$

and instead of $(\mathbf{g}_h - \mathbf{g}_i)'(\mathbf{g}_h - \mathbf{g}_i)$ being proportional to the Mahalanobis distance between $\mathbf{x}_h$ and $\mathbf{x}_i$ it is now equal to the Euclidean distance. This follows since

$$(\mathbf{x}_h - \mathbf{x}_i)'(\mathbf{x}_h - \mathbf{x}_i) = (\mathbf{g}_h - \mathbf{g}_i)'\mathbf{H}'\mathbf{H}(\mathbf{g}_h - \mathbf{g}_i)$$

$$= (\mathbf{g}_h - \mathbf{g}_i)'\mathbf{A}'\mathbf{A}(\mathbf{g}_h - \mathbf{g}_i)$$

$$= (\mathbf{g}_h - \mathbf{g}_i)'(\mathbf{g}_h - \mathbf{g}_i).$$

Therefore, if we prefer a plot on which the distance between $\mathbf{g}_h^*$ and $\mathbf{g}_i^*$ is a good approximation to Euclidean, rather than to Mahalanobis, distance between $\mathbf{x}_h$ and $\mathbf{x}_i$, then the biplot with $\alpha = 1$ will be preferred to $\alpha = 0$. Note that using Mahalanobis distance emphasizes the distance apart of the observations in the direction of the low-variance PCs, and downweights distances in the direction of high-variance PCs, when compared with Euclidean distance—see Section 10.1.

Another interesting property of the biplot with $\alpha = 1$ is that the positions of the $\mathbf{g}_i^*$'s are identical to those given by a straightforward plot with respect to the first two PCs, as described in Section 5.1. It follows from equation (5.3.3) and Section 3.5 that we can write

$$x_{ij} = \sum_{k=1}^{r} z_{ik} a_{jk},$$

where $z_{ik} = u_{ik} l_k^{1/2}$ is the value of the kth PC for the ith observation. But with $\alpha = 1$, $\mathbf{G} = \mathbf{UL}$, so the kth element of $\mathbf{g}_i$ is $u_{ik} l_k^{1/2} = z_{ik}$. $\mathbf{g}_i^*$ consists of the first two elements of $\mathbf{g}_i$, which are simply the values of the first two PCs for the ith observation.

The properties for the $\mathbf{h}_j$'s which were demonstrated above for $\alpha = 0$ will no longer be valid exactly for $\alpha = 1$, although similar interpretations can still be drawn, at least in qualitative terms. In fact, the co-ordinates of $\mathbf{h}_j^*$ are simply the coefficients of the jth variable for the first two PCs. The advantage of superimposing the plots of the $\mathbf{g}_i^*$'s and $\mathbf{h}_j^*$'s will remain for $\alpha = 1$, since

x_{ij} will still represent the projection of $\mathbf{g}_i$ onto $\mathbf{h}_j$. In many ways, the biplot with $\alpha = 1$ is nothing new, since the $\mathbf{g}_i^*$'s give PC scores, and $\mathbf{h}_j^*$'s give PC coefficients, both of which are widely used on their own. The biplot, however, superimposes both $\mathbf{g}_i^*$'s and $\mathbf{h}_j^*$'s to give additional information.

Other values of α could also be used; for example, Gabriel (1971) mentions $\alpha = \frac{1}{2}$, in which the sum of squares of the projections of plotted points onto either one of the axes, is the same for observations as for variables (Osmond, 1985), but most applications seem to have used $\alpha = 0$, or sometimes $\alpha = 1$. For other values of α the general qualitative interpretation of the relative positions of the $\mathbf{g}_i$'s and $\mathbf{h}_j$'s will remain the same, but the exact properties which hold for $\alpha = 0$ and $\alpha = 1$ will no longer be valid.

Another possibility is to superimpose $\mathbf{g}_i^*$'s and $\mathbf{h}_j^*$'s corresponding to different values of α. For example, we could use $\mathbf{g}_i^*$ corresponding to $\alpha = 1$ and $\mathbf{h}_j^*$ corresponding to $\alpha = 0$, so that the $\mathbf{g}_i^*$'s give a PC plot, and the $\mathbf{h}_j^*$'s have a direct interpretation in terms of variances and covariances. Mixing values of α in this way will, of course, lose the property that x_{ij} is the projection of $\mathbf{g}_i$ onto $\mathbf{h}_j$, but the relative positions of the $\mathbf{g}_i^*$'s and $\mathbf{h}_j^*$'s will still give qualitative information about the size of each variable for each observation.

Finally, the biplot can be adapted to cope with missing values (see Section 11.6) by introducing weights w_{ij} for each observation x_{ij}, when approximating x_{ij} by $\mathbf{g}_i^{*'}\mathbf{h}_j^*$. A weight of zero is given to missing values and a unit weight to those values which are present. The appropriate values for $\mathbf{g}_i^*$, $\mathbf{h}_j^*$ can be calculated using an algorithm which handles general weights, due to Gabriel and Zamir (1979).

5.3.1. Examples

Two examples are now presented illustrating the use of biplots. Many other examples have been given by Gabriel; in particular, see Gabriel (1981) which discusses seven examples. Another interesting recent example, which emphasizes the usefulness of the *simultaneous* display of both rows and columns of the data matrix, is given by Osmond (1985).

Artistic Qualities of Painters

In Figure 5.3. a biplot is given for the data set described earlier in the present chapter and consisting of four subjective measurements of artistic qualities for 54 painters. The plot given uses the adapted version of $\alpha = 0$ in preference to $\alpha = 1$, because with $\alpha = 1$ the points representing the four variables are all very close to the centre of the plot, leading to difficulties in interpretation. The co-ordinates of the 54 painters are therefore re-scaled versions of those displayed in Figure 5.2, but their relative positions are similar. For example, the group of three 'Seventeenth Century' painters at the bottom of the plot is

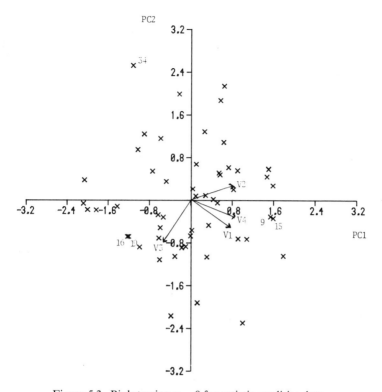

Figure 5.3. Biplot using $\alpha = 0$ for artistic qualities data.

still visible. Because of the expansion of the horizontal, relative to the vertical, scale the group of four painters at the left of the plot now seems to have been joined by a fifth, Murillo, who is from the same school as three of the others in this group. There is also an outlying painter, Fr. Penni, observation number 34, whose isolated position in the top left of the plot is perhaps more obvious on Figure 5.3 than Figure 5.2. The main distinguishing feature of this painter is that de Piles gave him a 0 score for composition, compared to a minimum of 4 (and maximum of 18) for all other painters.

Now consider the positions on the biplot of the vectors corresponding to the four variables. It is seen that composition and expression (V1 and V4) are close together, reflecting their relatively large positive correlation, and that drawing and colour (V2 and V3) are in opposite quadrants, confirming their fairly large negative correlation. Other correlations, and hence positions of vectors, are intermediate.

Finally, consider simultaneous positions of painters and variables. The two painters, numbered 9 and 15, which are slightly below the positive horizontal axis are Le Brun and Domenichino. These are close to the direction defined by V4, and not far from the directions of V1 and V2, which implies that they

Table 5.2. First two PCs: 100 km running data.

		Component 1	Component 2
first 10 km		−0.30	0.45
second 10 km		−0.30	0.45
third 10 km		−0.33	0.34
fourth 10 km		−0.34	0.20
fifth 10 km	Coefficients	−0.34	−0.06
sixth 10 km		−0.35	−0.16
seventh 10 km		−0.31	−0.27
eighth 10 km		−0.31	−0.30
ninth 10 km		−0.31	−0.29
tenth 10 km		−0.27	−0.40
Eigenvalue		7.24	1.28
Cumulative percentage of total variation		72.4	85.3

should have higher than average scores on these three variables. This is indeed the case—Le Brun scores 16 (on a scale from 0 to 20) on all three variables, and Domenichino scores 17 on V2 and V4 and 15 on V1. Their position relative to V3 suggests an average or lower score on this variable— the actual scores are 8 and 9, which confirms this suggestion. As another example consider the two painters, 16 and 19 (Giorgione and Da Udine), whose positions are virtually identical, in the bottom left-hand quadrant of Figure 5.3. These two painters have high scores on V3 (18 and 16) and below average scores on V1, V2 and V4. This behaviour, but with lower scores on V2 than on V1, V4, would be predicted from the points' positions on the biplot.

100 km Running Data

The second example consists of data on times taken for each of ten 10 km sections, by the 80 competitors who completed the Lincolnshire 100 km race in June 1984. There are thus 80 observations on ten variables. I am grateful to Ron Hindley, the race organizer, for distributing the results of the race in such a detailed form.

The variances and coefficients for the first two PCs, based on the correlation matrix for these data, are given in Table 5.2. Results for the covariance matrix are similar, though with higher coefficients in the first PC for the later sections of the race, since (means and) variances of the times taken for each section tend to increase later in the race. The first component measures the overall speed of the runners, and the second contrasts those runners who slow down substantially during the course of the race with those runners who maintain a more even pace. Together, the first two PCs account for more than 85% of the total variation in the data.

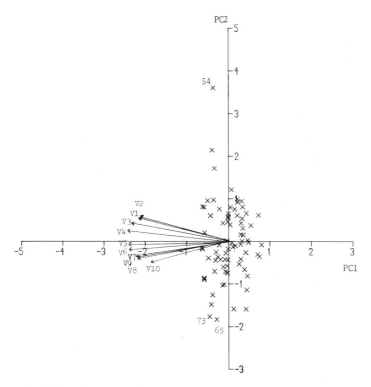

Figure 5.4. Biplot using $\alpha = 0$ for 100 km running data (V1, V2, ..., V10 indicate variables measuring times on first, second, ..., tenth sections of the race).

The adapted $\alpha = 0$ biplot for these data is shown in Figure 5.4. As with the previous example, the plot using $\alpha = 1$ is not very satisfactory because the vectors corresponding to the variables are all very close to the centre of the plot. Figure 5.4 shows that with $\alpha = 0$ we have the opposite extreme—the vectors corresponding to the variables and the points corresponding to the observations are completely separated. As a compromise, Figure 5.5 gives the biplot with $\alpha = \frac{1}{2}$, which at least has approximately the same degree of spread for variables and observations. As with $\alpha = 0$, the plot has been modified from the straightforward factorization corresponding to $\alpha = \frac{1}{2}$. The $\mathbf{g}_i$'s have been multiplied, and the $\mathbf{h}_j$'s divided, by $(n - 1)^{1/4}$, so that we have a compromise between $\alpha = 1$, and the adapted version of $\alpha = 0$. The adapted plot with $\alpha = \frac{1}{2}$ is still not entirely satisfactory since the vectors corresponding to variables all fall in a very narrow sector of the plot, but this is unavoidable for data which, as in the present case, have large correlations between all pairs of variables. The tight bunching of the vectors simply reflects large correlations, but it is interesting to note that the ordering of the vectors around their sector corresponds almost exactly to their position

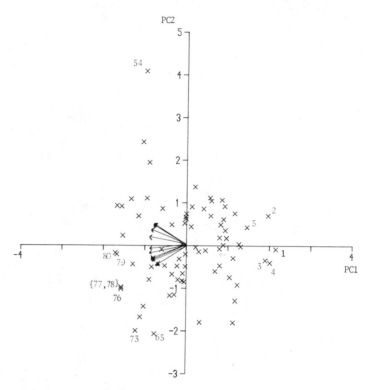

Figure 5.5. Biplot using $\alpha = \frac{1}{2}$ for 100 km running data (numbers indicate finishing position in race).

within the race. (The ordering is the same for both diagrams, but to avoid congestion, it has not been indicated on Figure 5.5.) With hindsight, this is not surprising since times in one part of the race are more likely to be similar to those in adjacent parts than to those which are more distant.

Turning to the positions of the observations, the points near the right-hand side of the diagrams correspond to the fastest athletes and those on the left to the slowest. To illustrate this, the first five and last five of the 80 finishers are indicated on Figure 5.5. (Note that competitors 77 and 78 ran together throughout the race—they therefore have identical values for all ten variables and PCs, and hence identical positions on the plot.) The positions of the athletes in this (horizontal) direction tally with the positions of the vectors—observations with large values on all variables (i.e. slow runners) will be in the direction of the vectors i.e. towards the left.

Similarly, the observations near the top of the diagram are athletes who maintained a fairly steady pace, while those at the bottom slowed down considerably during the race. Again this corresponds with the position of the vectors—those observations at the bottom of the diagram should have large values of V10, V9, V8, etc. compared with V1, V2, V3, etc. (i.e. they slow down a lot) whereas those at the top have nearly equal values for all variables. For example, consider the outlying observation at the top of Figures 5.4 and 5.5. This point corresponds to the 54th finisher who was the only competitor to run the final 10 km faster than the first 10 km. To put this into perspective it should be noted that the average times taken for the first and last 10 km by the 80 finishers were 47.6 min and 67.0 min respectively, so that most competitors slowed down considerably during the race.

At the opposite extreme to the 54th finisher, consider the two athletes corresponding to the points at the bottom left of the plots. These correspond to the 65th and 73rd finishers whose times for the first and last 10 km were 50.0 min and 87.8 min for the 65th finisher and 48.2 min and 110.0 min for the 73rd finisher. This latter athlete therefore ran at a nearly 'average' pace for the first 10 km but was easily one of the slowest competitors over the last 10 km.

5.4. Correspondence Analysis

The technique commonly called correspondence analysis has been 'redis-covered' many times in several different guises with various names—see Greenacre (1984), for a comprehensive treatment of the subject, and in partic-ular his Section 1.3 and Chapter 4 which discuss, respectively, the history and the various different approaches to the topic. The name 'correspondence analysis' is derived from the French 'analyses des correspondances' (Benzécri, 1980). Although, at first sight, correspondence analysis seems unrelated to PCA it can be shown that it is, in fact, equivalent to a form of PCA for discrete (generally nominal) variables (see Section 11.1). The technique is often used to provide a graphical representation, in two dimensions, of data which are presented in the form of a contingency table, and because of this graphical usage the technique is introduced briefly in the present chapter. Further discussion of correspondence analysis and various generalizations of the technique, together with its connections with PCA, is given in Sections 11.1, 12.1 and 12.2.

Suppose that a set of data is presented in the form of a two-way contin-gency table, in which a set of n observations is classified according to their values on two discrete random variables. Thus the information available is the set of frequencies $\{n_{ij}, i = 1, 2, \ldots, r; j = 1, 2, \ldots, c\}$ where n_{ij} is the number of observations which take the ith value for the first (row) variable and the jth value for the second (column) variable. Let $\mathbf{N}$ be the $(r \times c)$ matrix whose (i, j)th element is n_{ij}.

There are a number of seemingly different approaches which all lead to correspondence analysis; Greenacre (1984, Chapter 4) discusses these various possibilities in some detail. Whichever approach is used, the final product is a sequence of pairs of vectors $(\mathbf{f}_1, \mathbf{g}_1), (\mathbf{f}_2, \mathbf{g}_2), \ldots, (\mathbf{f}_q, \mathbf{g}_q)$ where $\mathbf{f}_k, k = 1, 2, \ldots$ are r-vectors of scores or coefficients for the rows of $\mathbf{N}$, and $\mathbf{g}_k, k = 1, 2, \ldots$ are c-vectors of scores or coefficients for the columns of $\mathbf{N}$. These pairs of vectors are such that the first q such pairs give a 'best-fitting' representation in q dimensions, in a sense to be defined in Section 11.1, of the matrix $\mathbf{N}$, and of its rows and columns. It is usual to take $q = 2$ and the rows and columns can then be plotted on a two-dimensional diagram; the co-ordinates of the ith row are the ith elements of $\mathbf{f}_1, \mathbf{f}_2, i = 1, 2, \ldots, r$, and the co-ordinates of the jth column are the jth elements of $\mathbf{g}_1, \mathbf{g}_2, j = 1, 2, \ldots, c$.

Such two-dimensional plots cannot in general be compared in any direct way with plots with respect to PCs, or biplots, since $\mathbf{N}$ is a different type of data matrix from that used in PCs or biplots. However, Greenacre (1984, Sections 9.6 and 9.10) also gives examples where correspondence analysis is done with an ordinary $(n \times p)$ data matrix, $\mathbf{X}$, replacing $\mathbf{N}$; this is only possible if all variables are measured in the same units. In these circumstances, correspondence analysis is producing a simultaneous two-dimensional plot of the rows and columns of $\mathbf{X}$, which is precisely what is done in a biplot, but the two analyses are not the same.

Both the biplot and correspondence analysis determine the plotting positions for rows and columns of $\mathbf{X}$ from the singular value decomposition (SVD) of a matrix—see Section 3.5. For the biplot, the SVD is calculated for the matrix $\mathbf{X}$ (after subtracting mean values $\bar{x}_j, j = 1, \ldots, p$ from the columns), but in correspondence analysis, the SVD is found for a matrix of *residuals*, after subtracting 'expected values assuming independence of rows and columns' from $(1/n_{\cdot})\mathbf{X}$ (see Section 11.1). The effect of looking at residual (or interaction) terms is (Greenacre, 1984, p. 288) that all the dimensions found by correspondence analysis represent aspects of the 'shape' of the data, whereas in PCA, the first PC often simply represents 'size' (see Section 4.1).

Correspondence analysis provides one way in which a data matrix may be adjusted, in order to eliminate some uninteresting feature, such as 'size', before finding a SVD and, hence, 'PCs'. Other possible adjustments, such as double-centring of $\mathbf{X}$, are discussed in Section 12.3.

5.4.1. Example

Figure 5.6 gives a plot obtained by correspondence analysis on a data set which recorded the presence or otherwise of 52 bird species at a number of wetland sites in Ireland. The data displayed in Figure 5.6 refer to summer sightings and are part of a much larger data set, which was kindly supplied by Dr. R. J. O'Connor of the British Trust for Ornithology, and which has been analysed in various ways in two unpublished student projects/dissertations (Worton, 1984; Denham, 1985) at the University of Kent. To avoid congestion on Figure 5.6, only a few of the points corresponding to sites and species have been labelled; these points will now be discussed. Although correspon-

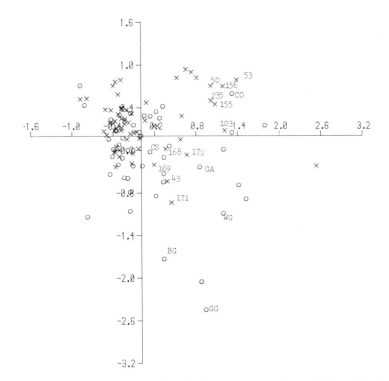

Figure 5.6. Correspondence analysis plot for summer species at Irish wetland sites. × denotes site; o denotes species.

dence analysis treats the data differently from a biplot, it is still true that sites, (or species) which are close to each other on the correspondence analysis plot are likely to be similar with respect to their values for the original data. Furthermore, as in a biplot, we can interpret the joint positions of sites and species.

On Figure 5.6 we first note that sites which are close to each other on the figure also tend to be close geographically. For example, the group of sites in the top-right of the plot {50, 53, 103, 155, 156, 235} are all inland sites in the south-west of Ireland, and the group {43, 168, 169, 171, 172} in the bottom-right of the diagram are all coastal sites in the south and east.

If we look at species, rather than sites, we find that similar species tend to be located in the same part of Figure 5.6. For example, three of the four species of goose which were recorded are in the bottom-right of the diagram (BG, WG, GG).

Turning to the simultaneous positions of species and sites, the Greylag Goose (GG) and Barnacle Goose (BG) were only recorded at site 171, among those sites which are numbered on Figure 5.6. Site 171 is closest in position, of any site, to the positions of these two species. The Whitefronted Goose

(WG) is recorded at sites 171 and 172 only, the Gadwall (GA) at sites 43, 103, 168, 169, 172 among those labelled on the diagram, and the Common Sandpiper (CS) at all sites in the coastal group {43, 168, 169, 171, 172}, but at only one of the inland group {50, 53, 103, 155, 156, 235}. Again, these occurrences might be predicted from the relative positions of the sites and species on the plot. However, simple predictions are not always valid, since the Coot (CO), whose position on the plot is in the middle of the inland sites, is recorded at all 11 sites numbered on the figure.

5.5. Comparisons Between Principal Co-ordinates, Biplots, Correspondence Analysis and Plots Based On Principal Components

For most purposes there is little point in asking which of the graphical techniques discussed so far in this chapter is 'best'. This is because they are either equivalent, as with PCs and principal co-ordinates for some types of similarity matrix, so any comparison is trivial, or the data set is of a type such that one or more of the techniques is not really appropriate, and so should not be compared with the others. For example, if the data are in the form of a contingency table, then correspondence analysis is clearly relevant, but the use of the other techniques is more questionable. The biplot is, in fact, not restricted to 'standard' $(n \times p)$ data matrices, and could be used on any two-way array of data, with the simultaneous positions of the g_i^*'s and h_j^*'s still having a similar interpretation to that discussed in Section 5.3, even though some of the separate properties of the g_i^*'s and h_j^*'s (for instance, relating to variances and covariances) are clearly no longer valid. A contingency table could also be analysed by 'PCA', but this is really not appropriate, since it is not at all clear what interpretation could be given to the results. Principal co-ordinate analysis needs a similarity or distance matrix, so it is hard to see how it could be properly used on a contingency table.

There are a number of connections between PCA and the other techniques— links with principal co-ordinate analysis and biplots have already been discussed, while those with correspondence analysis are deferred until Section 11.1—but for most data sets one of the methods is more appropriate than the others. Contingency table data require correspondence analysis, similarity or dissimilarity matrices imply principal co-ordinate analysis, whereas PCA is defined for 'standard' data matrices of n observations on p variables. Notwithstanding these distinctions, different techniques have been used on the same data sets and a number of empirical comparisons have been reported in the ecological literature. For example, Fasham (1977) and Gauch (1982), among others, have compared PCA with other techniques, including correspondence analysis, on simulated data. The data are generated to have a similar structure to that expected in some types of ecological data, with added noise, and investigations are conducted to see which techniques are 'best' at recovering the structure. However, as with comparisons between PCA and correspondence analysis given by Greenacre (1984, Section 9.6), the relevance of all the techniques compared, to the data analysed, is open to question. Different techniques implicitly assume that different types of structure or model are of interest for the data (see Section 12.3 for some further possibilities) and which technique is most appropriate will depend on which type of structure or model is relevant.

5.6. Methods for Graphical Display of Intrinsically High-Dimensional Data

Sometimes it will not be possible to reduce a data set's dimensionality to two or three without a substantial loss of information; in such cases, methods for displaying many variables simultaneously in two dimensions may be useful. Some of these methods are quite well known, such as the plots of trigonometric functions due to Andrews (1972), illustrated below, and the display in terms of faces suggested by Chernoff (1973), for which several examples are given in Wang (1978). There are, however, many other possibilities—see, for example, Tukey and Tukey (1981)—which will not be discussed here.

Even when dimensionality cannot be reduced to two or three, a reduction to as few dimensions as possible (without throwing away too much information) is still often worth while before attempting to graph the data. Some techniques, such as Chernoff's faces, impose a limit on the number of variables which can be handled (although a recent modification due to Flury and Riedwyl (1981) increases the limit), and for most other methods a reduction in the number of variables leads to simpler and more easily interpretable diagrams. An obvious way of reducing the dimensionality is to replace the original variables by the first few PCs, and the use of PCs in this context will be particularly successful if each PC retains an obvious interpretation (see Chapter 4). Andrews (1972), in fact, recommends transforming to PCs in any case, because the independence of the PCs means that tests of significance for the plots may be more easily performed with PCs than with the original variables.

5.6.1. Example

In Jolliffe *et al.* (1986), 107 English local authorities are divided into groups or clusters, using various methods of cluster analysis (see Section 9.2), on the basis of measurements on 20 demographic variables.

The 20 variables can be reduced to seven PCs (which account for over

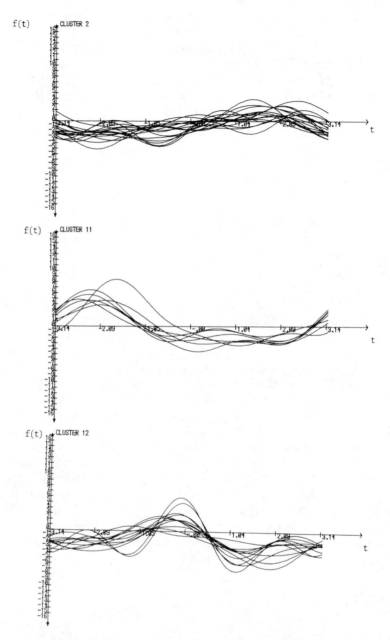

Figure 5.7. Local authorities demographic data: Andrews' curves for three clusters.

90% of the total variation in the 20 variables) and, for each individual, an Andrews' curve is defined on the range $-\pi \leq t \leq \pi$ by the function

$$f(t) = \frac{z_1}{\sqrt{2}} + z_2 \sin(t) + z_3 \cos(t) + z_4 \sin(2t) + z_5 \cos(2t)$$

$$+ z_6 \sin(3t) + z_7 \cos(3t),$$

where $z_1, z_2, \ldots, z_7$ are the values of the first seven PCs for the individual concerned. Andrews' curves may be plotted separately for each cluster, and these curves are useful in assessing the homogeneity of the clusters. For example, Figure 5.7 gives the Andrews' curves for three of the clusters (Clusters 2, 11 and 12) in a 13-cluster solution, and it can be seen immediately that the shape of the curves is different for different clusters. Compared to the variation between clusters, the curves fall into fairly narrow bands (with a few exceptions) for each cluster. Narrower bands for the curves imply greater homogeneity in the cluster.

In Cluster 12 there are two curves which are somewhat different from the remainder. These curves have three complete oscillations in the range $(-\pi, \pi)$, with maxima at 0, and $\pm 2\pi/3$. This implies that they are dominated by $\cos(3t)$, and hence z_7. Examination of the seventh PC shows that its largest coefficients are all positive, and correspond to numbers of elderly persons who have recently moved to the area, numbers in privately rented accommodation, and population sparsity (Area/Population). The implication of the outlying curves for Cluster 12 is that the two local authorities corresponding to the curves (Cumbria, Northumberland) have substantially larger values for the seventh PC than do the other local authorities in the same cluster (Cornwall, Gloucestershire, Kent, Lincolnshire, Norfolk, North Yorkshire, Shropshire, Somerset and Suffolk). This is, indeed, the case and it further implies atypical values for Northumbria and Cumbria, compared to the remainder of the cluster, for the three variables which have the largest coefficients for the seventh PC.

Another example of using Andrews' curves to examine the homogeneity of clusters in cluster analysis, and to investigate potential outliers, is given by Jolliffe et al. (1980), for the data set discussed in Section 4.2.

CHAPTER 6

Choosing a Subset of Principal Components or Variables

In this chapter two separate, but related, topics are considered, both of which are concerned with choosing a subset of variables. In the first section, the choice to be examined is how many PCs adequately account for the total variation in $\mathbf{x}$. The major objective in many applications of PCA is to replace the p elements of $\mathbf{x}$ by a much smaller number, m, of PCs, which nevertheless discard very little information. It is crucial to know how small m can be taken without serious information loss. Various rules, mostly *ad hoc*, have been proposed for determining a suitable value of m, and these are discussed in Section 6.1. Examples of their use are given in Section 6.2.

Using m PCs instead of p variables considerably reduces the dimensionality of the problem when $m \ll p$, but usually the values of all p variables are still needed in order to calculate the PCs, since each PC may be a function of all p variables. It might be preferable if, instead of using m PCs, we could use m (or perhaps slightly more) of the original variables, to account for most of the variation in $\mathbf{x}$. Again, several methods are available for selecting subsets of the p variables in such a way that most of the information available from the full set of variables is retained. Some of the methods use the PCs directly to decide upon a subset, and these are described in Section 6.3, together with a related class of techniques due to McCabe (1984). A brief description of two further techniques, which indirectly use the PCA to select a subset of variables completes Section 6.3, and Section 6.4 gives two examples of the use of variable selection methods.

All of the variable selection methods described in the present chapter are appropriate when the objective is to describe variation *within* $\mathbf{x}$ as well as possible. Variable selection when $\mathbf{x}$ is a set of regressor variables in a regression analysis, or a set of predictor variables in a discriminant analysis, is a different type of problem since criteria external to $\mathbf{x}$ must be considered. Variable selection in regression is discussed in Section 8.5.

6.1. How Many Principal Components?

In this section we present a number of rules for deciding how many PCs should be retained in order to account for most of the variation in **x** (or in the standardized variables, **x***, in the case of a correlation matrix).

In some circumstances the last few, rather than the first few, PCs are of interest, as was discussed in Section 3.4—see also Sections 3.7, 8.4, 8.6 and 10.1, and Section 6.3. In the present section, however, the traditional idea of trying to reduce dimensionality by replacing the p variables by the *first m* PCs $(m < p)$ is adopted, and the possible virtues of the last few PCs are ignored.

The first three types of rule for choosing m are very much *ad hoc* rules-of-thumb, whose justification, despite some attempts to put them on a more formal basis, is still mainly that they are intuitively plausible, and that they work in practice. The fourth is apparently statistically more respectable, but it is based on distributional assumptions which are often unrealistic, and, in any case, it seems to retain more variables than are necessary in practice. Three statistically based rules which do not require distributional assumptions are then described. Two of these methods use cross validation and the singular value decomposition (SVD), while the third considers partial correlations. Some procedures which have been suggested in the context of meteorology are then discussed briefly, and the section concludes with a few comments on the relative merits of various rules.

6.1.1. Cumulative Percentage of Total Variation

Perhaps the most obvious criterion for choosing m, which has already been informally adopted in some of the examples of Chapters 4 and 5, is to select a (cumulative) percentage of total variation which it is desired that the selected PCs should contribute, say 80% or 90%. The required number of PCs is then the smallest value of m for which this chosen percentage is exceeded. It remains to define what is meant by 'percentage of variation accounted for by the first k PCs', but this poses no real problem. Principal components are successively chosen to have the largest possible variance, and the variance of the jth PC is l_j. Furthermore, $\sum_{j=1}^{p} l_j = \sum_{j=1}^{p} s_{jj}$, that is the sum of the variances of the PCs is equal to the sum of the variances of the elements of **x**. The obvious definition of 'percentage of variation accounted for by the first k PCs' is therefore

$$t_k = 100 \sum_{j=1}^{k} l_j \bigg/ \sum_{j=1}^{p} s_{jj} = 100 \sum_{j=1}^{k} l_j \bigg/ \sum_{j=1}^{p} l_j,$$

which reduces to

$$t_k = \frac{100}{p} \sum_{j=1}^{k} l_j, \quad \text{in the case of a correlation matrix.}$$

Choosing a cut-off, t^*, somewhere between 70% and 90%, and retaining m PCs, where m is the smallest integer, k, for which $t_k > t^*$, provides a rule which, in practice, preserves in the first m PCs most of the information in $\mathbf{x}$. The best value for t^* will generally become smaller as p increases, or as n, the number of observations, increases.

Using the rule is, in a sense, equivalent to looking at the spectral decomposition of the covariance (or correlation) matrix, $\mathbf{S}$ (see Property A3 of Sections 2.1, 3.1), or the SVD of the data matrix, $\mathbf{X}$ (see Section 3.5). In either case, deciding how many terms to include in the decomposition in order to get a good fit to $\mathbf{S}$ or $\mathbf{X}$ respectively is closely related to looking at t_k, because an appropriate measure of goodness-of-fit of the first m terms in either decomposition is $\sum_{k=m+1}^{p} l_k$. This follows since

$$\sum_{i=1}^{n} \sum_{j=1}^{p} (_m\tilde{x}_{ij} - x_{ij})^2 = (n - 1) \sum_{k=m+1}^{p} l_k,$$

in the present notation (Gabriel, 1978) and $\| _m\mathbf{S} - \mathbf{S} \| = \sum_{k=m+1}^{p} l_k$ (see the discussion of Property G4 in Section 3.2), where $_m\tilde{x}_{ij}$ is the rank m approximation to x_{ij} based on the SVD, as given in equation (3.5.3), and $_m\mathbf{S}$ is the sum of the first m terms of the spectral decomposition of $\mathbf{S}$.

A number of attempts have been made to find the distribution of t_k, and hence to produce a formal procedure for choosing m, based on t_k. Mandel (1972) presents some expected values for t_k, for the case where all variables are independent, normally distributed, and have the same variance. Mandel's results are based on simulation studies, and although exact results have been produced by some authors, they are only for limited special cases. For example, Krzanowski (1979a) has given exact results for $k = 1$ and $p = 3$ or 4, again under the assumptions of normality, independence and equal variances for all variables. These assumptions mean that the results can be used to determine whether or not all variables are independent, but are of little general use in determining an 'optimal' cut-off for t_k. Sugiyama and Tong (1976) describe an approximate distribution for t_k which does not assume independence or equal variances, and which can be used to test whether $l_1, l_2, \ldots, l_k$ are compatible with any given structure for $\lambda_1, \lambda_2, \ldots, \lambda_k$, the corresponding population variances. However, the test still assumes normality and it is only approximate, so it is not clear how useful it will be in practice for choosing an appropriate value of m.

6.1.2. Size of Variances of Principal Components

The previous rule is equally valid whether a covariance or a correlation matrix is used to compute the PCs. The rule described in this section is constructed specially for use with correlation matrices, although it can be adapted for some types of covariance matrices. The idea behind the rule is that if all elements of $\mathbf{x}$ are independent, then the PCs are the same as the original variables, and all have unit variances in the case of a correlation matrix. Thus any PC with variance less than one contains less information than one of the original variables, and so is not worth retaining. The rule, in its simplest form, is sometimes called Kaiser's rule (Kaiser, 1960) and retains only those PCs whose variances, l_k, are ≥ 1. If the data set contains groups of variables which have large within-group correlations, but small between-group correlations, then there is one PC associated with each group whose variance is > 1, whereas any other PCs associated with the group have variances < 1 (see Section 3.8). Thus, the rule will generally retain one, and only one, PC associated with each such group of variables, which seems to be a reasonable course of action for data of this type.

As well as these intuitive justifications, Kaiser (1960) puts forward a number of other reasons for a cut-off at $l_k = 1$. It must be noted, however, that most of the reasons are pertinent to factor analysis (see Chapter 7), rather than PCA, although Kaiser refers to PCs in discussing one of them.

It can be argued that a cut-off at $l_k = 1$ retains too few variables. Consider a variable which, in the population, is more-or-less independent of all other variables. In a sample, such a variable will have small coefficients in $(p - 1)$ of the PCs but will dominate one of the PCs, whose variance, l_k, will be close to one when using the correlation matrix. Since the variable provides independent information from the other variables it would be unwise to delete it. However, deletion will occur if Kaiser's rule is used, and if, due to sampling variation, $l_k < 1$. It is therefore advisable to choose a cut-off, l^*, lower than one, to allow for sampling variation, and Jolliffe (1972) has suggested, based on simulation studies, that $l^* = 0.7$ is roughly the correct level. Further discussion of this cut-off level will be given with respect to examples in Sections 6.2 and 6.4.

The rule just described is specifically designed for correlation matrices, but it can be easily adapted for covariance matrices, by taking as a cut-off, l^*, the average value $\bar{l}$ of the eigenvalues, or, better, a somewhat lower cut-off such as $l^* = 0.7\bar{l}$. For covariance matrices with widely differing variances, however, this rule, and the one based on t_k from Section 6.1.1, retains very few (arguably, too few) PCs, as will be seen in the examples of Section 6.2.

An alternative way of looking at the sizes of individual variances is to use the so-called broken stick model. If we have a stick of unit length, which is broken, at random, into p segments, then it can be shown that the expected length of the kth longest segment is

$$l_k^* = \frac{1}{p} \sum_{j=k}^{p} \frac{1}{j}.$$

One way of deciding whether the proportion of variance accounted for by the kth PC is large enough for that component to be retained, is to compare the proportion with l_k^*. Principal components for which the proportion exceeds l_k^* are then retained, and all other PCs deleted. Tables of l_k^* are available for various values of p and k—see, for example, Legendre and Legendre (1983, p. 406).

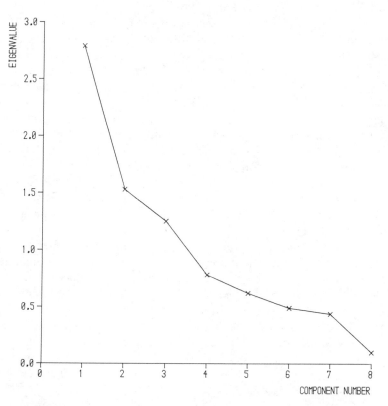

Figure 6.1. Scree graph for the correlation matrix: blood chemistry data.

6.1.3. The Scree Graph, and the Log–Eigenvalue Diagram

The first two rules described above involve a degree of subjectivity in the choice of cut-off levels, t^* and l^* respectively. The 'scree' graph, which was discussed and named by Cattell (1966), but which was already in common use, is even more subjective in its usual form, since it involves looking at a plot of l_k against k (see Figure 6.1, which is discussed in detail in Section 6.2) and deciding at which value of k the slopes of lines joining the plotted points are 'steep' to the left of k, and 'not steep' to the right. This value of k is then taken to be the number of components, m, to be retained. An alternative to the scree graph, which is popular in meteorology, is to plot $\log(l_k)$, rather than l_k, against k; this is known as the log-eigenvalue (or LEV) diagram—see Farmer (1971), Maryon (1979).

In introducing the scree graph, Cattell (1966) gives a somewhat different formulation from that above, and presents strong arguments that when it is used in factor analysis it is entirely objective and should produce a 'correct' number of factors. In fact, Cattell (1966) views the rule as a means of deciding upon an upper bound to the true number of factors in a factor analysis after rotation (see Chapter 7). He did not seem to envisage its use in PCA, although it has certainly been widely adopted for that purpose.

The way in which Cattell (1966) formulates the rule is to look for the point beyond which the scree graph defines a more-or-less straight line, not necessarily horizontal. The first point on the straight line is then taken to be the last factor/component to be retained. If there are two or more straight lines formed by the lower eigenvalues, then the cut-off is taken at the upper (left-hand) end of the left-most straight line.

The rule in Section 6.1.1 is based on $t_k = \sum_{j=1}^k l_j$, that in Section 6.1.2 looks at individual eigenvalues l_k, and the current rule, as applied to PCA, uses $l_{k-1} - l_k$ as its criterion. There is, however, no formal numerical cut-off based on $l_{k-1} - l_k$, and, in fact, judgements of when $l_{k-1} - l_k$ stops being large (steep) will depend on the *relative* values of $l_{k-1} - l_k$ and $l_k - l_{k+1}$, as well as the *absolute* value of $l_{k-1} - l_k$. Thus the rule is based subjectively on the second, as well as first, differences among the l_k's. Because of this, it is difficult to write down a formal numerical rule and the procedure has remained purely graphical.

Cattell's formulation, where we look for the point at which $l_{k-1} - l_k$ becomes fairly constant for several subsequent values is perhaps less subjective, but still requires some degree of judgement. Both formulations of the rule seem to work well in practice, provided that there is a fairly sharp 'elbow', or change in slope, in the graph. However, if the slope gradually becomes less steep, with no clear elbow, as in Figure 6.1, then it is clearly less easy to use the procedure.

Turning to the LEV diagram, an example of which is given in Section 6.2.2 below, one of the earliest published descriptions was in Craddock and Flood (1969), although, like the scree graph, it had been used routinely for some time before this. Craddock and Flood (1969) argued that, in meteorology, eigenvalues corresponding to 'noise' should decay in a geometric progression, and such eigenvalues will therefore appear as a straight line on the LEV diagram. Thus, to decide on how many PCs to retain, we should look for a point beyond which the LEV diagram becomes, approximately, a straight line. This is the same procedure as in Cattell's interpretation of the scree graph, but the results are different, since we are now plotting $\log(l_k)$ rather than l_k. To justify Craddock and Flood's procedure, Farmer (1971) generated simulated data with various known structures (or no structure). For purely random data, with all variables uncorrelated, Farmer found that the whole of the LEV diagram is approximately a straight line. Furthermore, he showed that if structures of various dimensions were introduced, then the LEV diagram was useful in indicating the correct dimensionality, although real examples, or course, give much less clear-cut results than those of simulated data.

6.1.4. The Number of Components with Unequal Eigenvalues

In Section 3.7.3 a test was described for the null hypothesis H_{0q}: $\lambda_{q+1} = \lambda_{q+2} = \cdots = \lambda_p$ against the general alternative that at least two of the last $p - q$ eigenvalues are unequal. It was argued that using this test for various values of q, it could be discovered how many of the PCs contributed substantial amounts of variation, and how many were simply 'noise'. If m, the required number of PCs to be retained, is defined as the number which are not noise, then the test can be used sequentially to find m.

$H_{0,p-2}$ is tested first, i.e. $\lambda_{p-1} = \lambda_p$ and if $H_{0,p-2}$ is accepted, then $H_{0,p-3}$ is tested. If $H_{0,p-3}$ is accepted, $H_{0,p-4}$ is tested, and this sequence continues until $H_{0,q}$ is first rejected at $q = q^*$, say. m is then taken to be $q^* + 1$ (or $q^* + 2$ if $q^* = p - 2$). There are a number of disadvantages to this procedure, the first of which is that equation (3.7.6) is based on the assumption of multivariate normality for $\mathbf{x}$, and is only approximately true even then. The second problem is concerned with the fact that unless $H_{0,p-2}$ is rejected, there are several tests to be done, so that the overall significance level of the sequence of tests is not the same as the individual significance levels of each test. Furthermore, it is difficult to get even an approximate idea of the overall significance level because the number of tests done is random and not fixed, and the tests are not independent of each other. It follows that, although the testing sequence suggested above can be used to estimate m, it is dangerous to treat the procedure as a respectable piece of statistical inference, since significance levels are almost completely unknown.

The procedure could be added to the list of *ad hoc* rules, but it has one further, more practical, disadvantage, namely that, in nearly all real examples, it tends to retain more PCs than are really necessary. Bartlett (1950), in introducing the procedure for correlation matrices, refers to it as testing how many of the PCs are statistically significant, but 'statistical significance' in the context of these tests does not imply that a PC accounts for a substantial proportion of the total variation. For correlation matrices, Jolliffe (1970) found that the rule often corresponds roughly to choosing a cut-off l^* of about 0.1 to 0.2 in the method of Section 6.1.2. This is much smaller than is recommended in that section, and occurs because defining unimportant PCs as those with variances equal to that of the last PC is not necessarily a sensible way of finding m. If this definition is acceptable, then the sequential testing procedure gives satisfactory results, but it is easy to construct examples where the method will give silly answers. For instance, if there is one near-constant relationship among the elements of $\mathbf{x}$, with a much smaller variance than any other PC, then the procedure will reject $H_{0,p-2}$ and declare that all PCs need to be retained, regardless of how nearly equal are the next few eigenvalues.

The method of this section is similar in spirit to, though more formalized than, one formulation of the scree graph. Looking for the first 'shallow' slope in the graph corresponds to looking for the first of two consecutive eigenvalues which are nearly equal. The scree graph differs from the formal testing procedure in that it starts from the largest eigenvalue, and compares consecutive eigenvalues two at a time, whereas the tests start with the smallest eigenvalues and compare blocks of two, three, four and so on. The scree graph is also more subjective, but, as has been stated above, the objectivity of the testing procedure is something of an illusion.

Cattell's original formulation of the scree graph differs from the above since it is differences $l_{k-1} - l_k$, rather than l_k, which must be equal beyond the cut-off point.

6.1.5. Cross Validatory Choice of m

It was noted in Section 6.1.1 that the rule described there is equivalent to looking at how well the data matrix $\mathbf{X}$ is fitted by the rank m approximation based on the SVD. The idea behind the methods discussed in the present section is similar, except that each element, x_{ij}, of $\mathbf{X}$ is now predicted from an equation like the SVD, but based on a submatrix of $\mathbf{X}$ which does not include x_{ij}. Two such methods have been suggested by Wold (1978) and by Eastment and Krzanowski (1982). In each, the number of terms in the estimate for $\mathbf{X}$, corresponding to the number of PCs, is successively taken as $1, 2, \ldots$, and so on, until overall prediction of the x_{ij}'s is no longer significantly improved by the addition of extra terms (PCs). The number of PCs to be retained, m, is then taken to be the minimum number necessary for adequate prediction.

Using the SVD, x_{ij} can be written, as in equations (3.5.2), (5.3.3),

$$x_{ij} = \sum_{k=1}^{r} u_{ik} l_k^{1/2} a_{jk},$$

where r is the rank of $\mathbf{X}$. [Recall that, in this context, l_k, $k = 1, 2, \ldots, p$ are eigenvalues of $\mathbf{X}'\mathbf{X}$, rather than of $\mathbf{S}$.]

An estimate of x_{ij}, based on the first m PCs, using all the data, is

$$_m\tilde{x}_{ij} = \sum_{k=1}^{m} u_{ik} l_k^{1/2} a_{jk}, \tag{6.1.1}$$

but what is required is an estimate based on a subset of the data which does not include x_{ij}. This estimate is written

$$_m\hat{x}_{ij} = \sum_{k=1}^{m} \hat{u}_{ik} \hat{l}_k^{1/2} \hat{a}_{jk}, \tag{6.1.2}$$

where $\hat{u}_{ik}$, $\hat{l}_k$, $\hat{a}_{jk}$ are estimated from a suitable subset of the data. The sum of squared differences between predicted and observed x_{ij}'s is then

$$\text{PRESS}(m) = \sum_{i=1}^{n} \sum_{j=1}^{p} (_m\hat{x}_{ij} - x_{ij})^2. \tag{6.1.3}$$

The notation PRESS, stands for PREdiction Sum of Squares, and is taken from the similar concept in regression, due to Allen (1974). All of the above is essentially common to both Wold (1978) and Eastment and Krzanowski (1982); they differ in how a subset is chosen for predicting x_{ij}, and in how (6.1.3) is used for deciding on m.

Eastment and Krzanowski (1982) use an estimate $\hat{a}_{jk}$ in (6.1.2) based on the data set with just the ith observation $\mathbf{x}_i$ deleted, $\hat{u}_{ik}$ is calculated with only the jth variable deleted, and $\hat{l}_k$ combines information from the two cases, with the ith observation and the jth variable deleted respectively. Wold (1978), on the other hand, divides the data into g blocks, where he recommends that g should be between four and seven and must not be a divisor of p, and that no block should contain the majority of the elements in any row or column of $\mathbf{X}$. Quantities equivalent to $\hat{u}_{ik}$, $\hat{l}_k$ and $\hat{a}_{jk}$ are calculated g times, once with each block of data deleted, and the estimates formed with the hth block deleted are then used to predict the data in the hth block, $h = 1, 2, \ldots, g$.

With respect to the choice of m, Wold (1978) and Eastment and Krzanowski (1982) each use a (different) function of PRESS(m) as a criterion for choosing m. To decide on whether to include the mth PC, Wold (1978) examines the ratio

$$R = \frac{\text{PRESS}(m)}{\sum_{i=1}^n \sum_{j=1}^p (_{m-1}\tilde{x}_{ij} - x_{ij})^2}. \tag{6.1.4}$$

This compares the prediction error sum-of-squares after fitting m components, with the sum of squared differences between observed and estimated data points based on all the data, using $(m - 1)$ components. If $R < 1$, then the implication is that a better prediction is achieved using m, rather than $(m - 1)$, PCs, so that the mth PC should be included.

The approach of Eastment and Krzanowski (1982) is similar to that in an analysis of variance. The reduction in prediction (residual) sum-of-squares in adding the mth PC to the model, divided by its degrees of freedom, is compared to the prediction sum-of-squares after fitting m PCs, divided by its degrees of freedom. Their criterion is thus

$$W = \frac{[\text{PRESS}(m - 1) - \text{PRESS}(m)]/v_{m,1}}{\text{PRESS}(m)/v_{m,2}}, \tag{6.1.5}$$

where $v_{m,1}, v_{m,2}$ are the degrees of freedom associated with the numerator and denominator respectively. It is suggested that if $W > 1$, then inclusion of the mth PC is worthwhile, although this cut-off at unity is to be interpreted with some flexibility. It is certainly not appropriate to stop adding PCs as soon as (6.1.5) first falls below unity, because the criterion is not necessarily a monotonic decreasing function of m.

It should be noted that although the criteria described in this section are somewhat less *ad hoc* than those of Sections 6.1.1–6.1.3, there is still no real attempt to set up a formal significance test to decide on m. Some progress has been made by Krzanowski (1983) in investigating the sampling distribution of W, using simulated data. He points out that there are two sources of variability to be considered in constructing such a distribution, namely the variability due to different sample covariance matrices, S, for a fixed population covariance matrix, Σ, and the variability due to the fact that a fixed sample covariance matrix S can result from different data matrices X. In addition to this two-tiered variability, there are a large number of parameters which can vary: n, p, and, particularly, the structure of Σ. This means that simulation studies can only examine a fraction of the possible parameter values, and will therefore be of restricted applicability. Krzanowski (1983) looks at several different types of structure for Σ, and reaches the conclusion that W chooses about the right number of PCs in each case, although there is a tendency for m to be too small. Wold (1978) also found, in a small simulation study, that R retained too few PCs. This underestimation for m could clearly be overcome by moving the cut-offs of W and R, respectively, slightly below and slightly above unity.

Although the cut-offs at $R = 1$ and $W = 1$ seem sensible, the reasoning behind them is not rigid, and they could be relaxed slightly to account for sampling variation, in the same way that Kaiser's rule (Section 6.1.2) seems to work better when l^* is changed to a value somewhat below unity.

6.1.6. Partial Correlation

For a PCA on the correlation matrix, Velicer (1976) suggests that the partial correlations between the p variables, given the values of the first m PCs, may be used to determine how many PCs to retain. The criterion proposed is the average of the squared partial correlations,

$$V = \sum_{\substack{i=1 \\ i \neq j}}^{p} \sum_{j=1}^{p} (r_{ij}^*)^2 / p(p-1),$$

where r_{ij}^* is the partial correlation between the ith and jth variables, given the first m PCs. r_{ij}^* is defined as the correlation between the residuals from the linear regression of the ith variable on the first m PCs, and the residuals from the corresponding regression of the jth variable on the m PCs. It therefore measures the strength of the linear relationship between the ith and jth variables, after removing the common effect of the first m PCs.

The criterion, V, will first decrease, and then increase, as m increases, and Velicer (1976) suggests that the optimal value of m corresponds to the minimum value of the criterion. Unfortunately, although this criterion is reasonable as a means of deciding the number of factors in a factor analysis (see Chapter 7), it is inappropriate in PCA. This is because it will not retain PCs which are dominated by a single variable, whose correlations with all the other variables are close to zero. Such variables are generally omitted from a factor model, but they provide information not available from other variables, so that they should be retained if most of the information in x is to be kept. It is for the same reason that Kaiser's rule should be modified when used in PCA, although in its original form it is satisfactory for factor analysis.

6.1.7. Rules for a Meteorological Context

As mentioned in Section 4.3, PCA has been widely used in meteorology and climatology to summarize data which vary both spatially and temporally, and a number of rules for selecting a subset of PCs have been put forward with this meteorological context very much in mind. The LEV diagram, discussed in Section 6.1.3, is one example, but there are many others. In particular, Preisendorfer (1981) suggests, because the different observations correspond to different time points, that important PCs will be those for which there is a clear pattern, rather than pure randomness, present in their behaviour through time. The important PCs can then be discovered by forming a time series of each PC, and testing which time series are distinguishable from white noise. Many tests are available for this purpose in time series literature, and Preisendorfer (1981) proposes the use of a number of them. This type of test is perhaps relevant in cases where the set of multivariate observations form a time series (see Section 11.2), as in many meteorological applications, but in the more usual (non-meteorological) situation where the observations are independent, such techniques are irrelevant, since the values of the PCs for different observations will also be independent. All PCs should therefore look like white noise.

Preisendorfer (1981) also discusses a number of other rules, which seem to be similar in spirit to the rules of Sections 6.1.3 and 6.1.4 above. They are, however, derived from consideration of a physical model, based on spring-coupled masses, where it is required to distinguish signal (the important PCs) from noise (the unimportant PCs). The details of the rules are, as a consequence, somewhat different from those of Sections 6.1.3 and 6.1.4, and they may also be of limited use outside the particular models studied. One of the methods suggested by Preisendorfer (1981)—see also Overland and Preisendorfer (1982)—involves simulating a large number of uncorrelated sets of data of the same size as the real data set which is to be analysed, and computing the eigenvalues of each simulated data set. To assess the significance of the eigenvalues for the real data set, they are compared to percentiles derived empirically from the simulated data. The idea of using simulated data to assess significance of eigenvalues has also been explored by other authors, for example Farmer (1971)—see also Section 6.1.3 above—Cahalan (1983), and, outside the meteorological context, Mandel (1972).

Other methods have also been suggested in the meteorological literature. For example, Jones et al. (1983) use a criterion, for correlation matrices, which was apparently devised by Guiot (1981). In this method PCs are retained if their cumulative eigenvalue product exceeds one. This technique will retain more PCs than most of the other procedures discussed earlier, but

Jones *et al.* (1983) seem to be satisfied with the results which it produces. Preisendorfer and Mobley (1982, Part IV) suggest a rule which considers retaining subsets of *m* PCs which are not necessarily restricted to the first *m*. This is reasonable if the PCs are to be used for an external purpose, such as regression (see Chapter 8), but is not really relevant if we are merely interested in accounting for as much of the variation in **x** as possible.

6.1.8. Discussion

Although a large number of rules has been examined in the last seven subsections, the list is by no means exhaustive. For example, in Section 5.1 it was noted that superimposing a minimum spanning tree on a plot of the observations with respect to the first two PCs, will give a subjective indication of whether or not a *two*-dimensional representation is adequate. It is not possible to give *definitive* guidance on which rules are best, but we conclude this section with a few comments on their relative merits.

Some procedures, such as those introduced in Sections 6.1.4 and 6.1.6, are inappropriate in most circumstances, because they retain, respectively, too many or too few PCs, in general. Some rules have been derived in particular fields of application, such as meteorology (Sections 6.1.3, 6.1.7) or psychology (Sections 6.1.3, 6.1.6) and may be less relevant outside these fields than within them. The simple rules of Sections 6.1.1 and 6.1.2, seem to work well in many examples, although the recommended cut-offs must be treated flexibly, and may change somewhat depending on the values on the values on *n* and *p*, and on the presence of variables with dominant variances—see the examples in the next section. Attempts which have so far been made to construct rules which have more sound statistical foundations (Sections 6.1.1, 6.1.4, 6.1.5) seem, at present, to offer little advantage over the simpler rules.

6.2. Choosing *m*, the Number of Components: Examples

Two examples, will be given here to illustrate several of the techniques described in Section 6.1; in addition, the examples of Section 6.4 include some relevant discussion.

6.2.1. Clinical Trials Blood Chemistry

These data were introduced in Section 3.3, and consist of measurements of eight blood chemistry variables on 72 patients. The eigenvalues for the correlation matrix are given in Table 6.1, together with the related informa-

Table 6.1. First six eigenvalues for the correlation matrix, blood chemistry data.

Component number, k	1	2	3	4	5	6
Eigenvalue, l_k	2.79	1.53	1.25	0.78	0.62	0.49
$t_k = 100 \sum_{j=1}^{k} l_j/p$	34.9	54.1	69.7	79.4	87.2	93.3
$l_{k-1} - l_k$		1.26	0.28	0.47	0.16	0.13

tion which is required to implement the three *ad hoc* methods described in Sections 6.1.1–6.1.3.

Looking at Table 6.1 and Figure 6.1, the three *ad hoc* methods of Sections 6.1.1–6.1.3 suggest that between three and six PCs should be retained, but the decision on a single best number is not clear-cut. Four PCs account for nearly 80% of the total variation, but it takes six PCs to account for 90%. A cut-off at $l^* = 0.7$ for the second criterion retains four PCs, but the next eigenvalue is not very much smaller, so perhaps five should be retained. In the scree graph the slope actually increases between $k = 3$ and 4, but then falls sharply and levels off, suggesting that perhaps only four PCs should be retained. The LEV diagram (not shown) is of little help here; it has no clear indication of constant slope after any value of k, and has, in fact, its steepest slope between $k = 7$ and 8.

Using Cattell's (1966) formulation, there is no strong straight-line behaviour after any particular point, although perhaps a cut-off at $k = 4$ is most appropriate. Cattell suggests that the first point on the straight line (that is, the 'elbow' point) should be retained. However, if we consider the scree graph in the same light as the test of Section 6.1.4, then all eigenvalues after, and including, the elbow are deemed roughly equal and so all corresponding PCs should be deleted. This would lead to the retention of only three PCs in the present case.

Turning to Table 6.2, which gives information for the covariance matrix, corresponding to that which was presented for the correlation matrix in Table 6.1, the three *ad hoc* measures all conclusively suggest that one PC is sufficient, since the first PC accounts for such an overwhelming proportion of the total variation. It is undoubtedly true that choosing $m = 1$ accounts for

Table 6.2. First six eigenvalues for the covariance matrix, blood chemistry data.

Component number, k	1	2	3	4	5	6
Eigenvalue, l_k	1704.68	15.07	6.98	2.64	0.13	0.07
$l_k/\bar{l}$	7.88	0.07	0.03	0.01	0.0006	0.0003
$t_k = 100 \sum_{j=1}^{k} l_j \big/ \sum_{j=1}^{p} l_j$	98.6	99.4	99.8	99.99	99.995	99.9994
$l_{k-1} - l_k$		1689.61	8.09	4.34	2.51	0.06

the vast majority of the variation in **x**, but this conclusion is not particularly informative since it merely reflects that one of the original variables accounts for nearly all the variation in **x**. The PCs for the covariance matrix in this example were discussed further in Section 3.3, and it can be argued that the use of the covariance matrix, rather than the rules of Sections 6.1.1–6.1.3, is inappropriate for these data.

6.2.2. Gas Chromatography Data

These data, which were originally presented by McReynolds (1970), and which have been analysed by Wold (1978) and by Eastment and Krzanowski (1982), are concerned with gas chromatography retention indices. After removal of a number of apparent outliers and an observation with a missing value, there remain 212 (Eastment and Krzanowski) or 213 (Wold) measurements on ten variables. Wold (1978) claims that his method indicates the inclusion of five PCs in this example but, in fact, he strengthens slightly his criterion for retaining PCs. His nominal cut-off for including the kth PC is $R < 1$; the sixth PC has $R = 0.99$ (see Table 6.3) but he nevertheless chooses to exclude it. Eastment and Krzanowski (1982) also modify their nominal cut-off but in the opposite direction, so that an *extra* PC is included. The values of W for the third, fourth and fifth PCs are 1.90, 0.92, 0.41 (see Table 6.3) so the formal rule, excluding PCs with $W < 1$, would retain three PCs. However, because the value of W is fairly close to unity, Eastment and Krzanowski (1982) suggest that it is reasonable to retain the fourth PC as well.

It is interesting to note that this example is based on a covariance matrix, and has a very similar structure to that of the previous example when the covariance matrix was used. Information for the present example, corresponding to Table 6.2, is given in Table 6.3, for 212 observations. Also given in Table 6.3 are Wold's R (for 213 observations) and Eastment and Krzanowski's W.

It can be seen from Table 6.3, as with Table 6.2, that the first two of the *ad*

Table 6.3. First six eigenvalues for the covariance matrix, gas chromatography data.

Component number, k	1	2	3	4	5	6
Eigenvalue, l_k	312187	2100	768	336	190	149
$l_k/\bar{l}$	9.88	0.067	0.024	0.011	0.006	0.005
$t_k = 100 \sum_{j=1}^{k} l_j \Big/ \sum_{j=1}^{p} l_j$	98.8	99.5	99.7	99.8	99.9	99.94
$l_{k-1} - l_k$		310087	1332	432	146	51
R	0.02	0.43	0.60	0.70	0.83	0.99
W	494.98	4.95	1.90	0.92	0.41	0.54

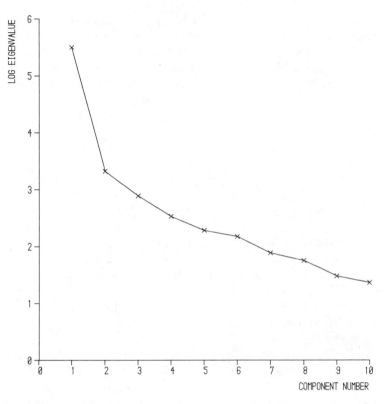

Figure 6.2. LEV diagram for the covariance matrix: gas chromatography data.

hoc methods will retain only one PC. The scree graph, which cannot be sensibly drawn because $l_1 \gg l_2$, would be more equivocal; it is clear from Table 6.3 that the slope drops very sharply after $k = 2$, indicating $m = 2$ (or 1), but each of the slopes for $k = 3, 4, 5, 6$ is substantially smaller than the previous slope, with no obvious levelling off. Nor is there any suggestion, for any cut-off, that the later eigenvalues lie on a straight line. There is, however, an indication of a straight line, starting at $m = 4$, in the LEV plot, which is given in Figure 6.2.

It would seem, therefore, that the cross-validatory criteria, R and W, differ considerably from the *ad hoc* rules (except perhaps the LEV plot) in the way in which they deal with covariance matrices which include a very dominant PC. Whereas most of the *ad hoc* rules will invariably retain only one PC in such situations, the present example shows that the cross-validatory criteria may retain several more. Krzanowski (1983) suggests that W looks for large gaps among the ordered eigenvalues, which is a similar aim to that of the scree graph, and that W can, therefore, be viewed as an objective analogue of the scree diagram. However, although this interpretation may be valid for the correlation matrices in his simulations, it does not seem to hold for the dominant variance structures exhibited in Tables 6.2 and 6.3.

For correlation matrices, and presumably for covariance matrices with less extreme variation among eigenvalues, the *ad hoc* methods and the cross-validatory criteria are likely to give more similar results. This is illustrated by a simulation study in Krzanowski (1983), where W is compared with the first two *ad hoc* rules with cut-offs at $t^* = 75\%$ and $l^* = 1$ respectively. The hypothesis testing procedure described in Section 6.1.4 is also included in the comparison but, as expected from the earlier discussion, it retains too many PCs in most circumstances. The behaviour of W compared with the two *ad hoc* rules, is the reverse of that observed in the example above. W retains fewer PCs than the $t_k \geq 75\%$ criterion, despite the fairly low cut-off of 75%. Similar numbers of PCs are retained for W and for the rule based on $l_k \geq 1$. The latter rule will retain more PCs if the cut-off is lowered to 0.7, rather than 1.0, as suggested in Section 6.1.2. It can also be argued that the cut-off for W should be reduced below unity, in which case all three rules will give similar results.

Krzanowski (1983) also examines the gas chromatography example further by generating six different artificial data sets with the same sample covariance matrix as the real data. The values of W are fairly stable across the replicates and confirm the choice of four PCs, obtained above by slightly decreasing the cut-off for W. For the full data set, with outliers not removed, the replicates give some different, and useful, information from that in the original data.

6.3. Selecting a Subset of Variables

When p, the number of variables observed, is large it is often the case that a subset of m variables, with $m \ll p$, will contain virtually all the information available in all p variables. It would be useful to determine an appropriate value of m, and then to decide which subset or subsets of m variables are best.

Solution of these two problems, the choice of m and the selection of a good subset, depends on the purpose to which the subset of variables is to be put. If the purpose is simply to preserve most of the variation in **x**, then the PCs of **x** can be used fairly straightforwardly to solve both problems, as will be explained shortly. A more familiar variable selection problem is in multiple regression, and although PCA can be useful in this context, it is used in a more complicated manner. This is because external considerations (the relationships of the predictor (regressor) variables with the dependent variable), as well as the internal relationships between the regressor variables, must be considered. External considerations are also relevant in other variable selection situations, such as in discriminant analysis (Section 9.1); such situations will not be considered in the present chapter. Furthermore, practical considerations, such as ease of measurement of the selected variables, may be important in some circumstances, and it must be stressed that such considerations, as well as the purpose of the subsequent analysis, may play a prominent role in variable selection. Here, however, we concentrate on the problem of finding a subset of **x** in which the sole aim is to represent the internal variation of **x** as well as possible. Variable selection in regression will be discussed in Section 8.5.

Regarding the choice of m, the methods of Section 6.1 are all relevant. The techniques described there find the number of PCs which account for most of the variation in **x**, but they can also be interpreted as finding the effective dimensionality of **x**. If **x** can be successfully described by only m PCs, then it will often be true that **x** can be replaced by a subset of m (or perhaps slightly more) variables, with a relatively small loss of information.

Moving on to the choice of m variables, Jolliffe (1970, 1972, 1973) discussed a number of methods for selecting a subset of m variables which preserve most of the variation in **x**. Some of the methods which were compared, and indeed some of those which performed quite well, were based on PCs. Other methods, including some based on cluster analyses of variables (see Section 9.2) were also examined, but, since these did not use the PCs to select variables, they are not described here. Three main types of method using PCs were examined.

(i) Associate one variable with each of the last $m_1^* = p - m_1$ PCs and delete those m_1^* variables. This can either be done once only, or iteratively. In the latter case a second PCA is done on the m_1 remaining variables, and a further set of m_2^* variables is deleted, if appropriate. A third PCA can then be done on the $p - m_1^* - m_2^*$ variables, and the procedure is repeated until no further deletions are considered necessary. $m_1^*, m_2^*, \ldots,$ are chosen by a criterion based on size of eigenvalues l_k.

The reasoning behind this method is that small eigenvalues correspond to near-constant relationships between a subset of variables. If one of the variables involved in such a relationship is deleted (a sensible choice for deletion is the variable with the highest coefficient in absolute value, in the relevant PC), little information is lost. To decide on how many variables to delete, the obvious criterion is l_k, as described in Section 6.1.2. The criterion, t_k, of Section 6.1.1 was also tried by Jolliffe (1972) but shown to be less useful.

(ii) Associate a set of m^* variables, *en bloc*, with the last m^* PCs, and then delete these variables. Jolliffe (1970, 1972) investigated this type of method, with the m^* variables either chosen to maximize sums of squares of coefficients in the last m^* PCs, or to be those m^* variables which are best predicted by regression on the first $m = p - m^*$ PCs. Choice of m^* is again based on the sizes of the l_k's. It was found that such methods were unsatisfactory, since they consistently failed to select an appropriate subset for some simple correlation structures.

(iii) Associate one variable with each of the first m PCs, namely the variable not already chosen, with the highest coefficient, in absolute value, in each successive PC. These m variables are retained, and the remaining $m^* = p - m$ are deleted.

The arguments leading to this approach are twofold. First, it is an obvious complementary approach to (i), and, second, in cases where there are groups of highly correlated variables it is designed to select just one variable from each group. This will happen because there will be exactly one high-variance PC associated with each group (see Section 3.8). The approach is a plausible one, since a single variable from each group should preserve most of the information given by that group, when all variables in the group are highly correlated.

In Jolliffe (1972) comparisons were made, using simulated data, between a non-iterative version of method (i) and method (iii), called methods B2, B4 respectively, and with several other subset selection methods which do not use the PCs. The results showed that the PC methods B2, B4 retained the 'best' subsets more often than the other methods considered, but they also selected 'bad' as opposed to 'good' or 'moderate', subsets more frequently than the other methods. Method B4 was most extreme in this respect; it selected 'best' and 'bad' subsets more frequently than any other method, and 'moderate' or 'good' subsets less frequently.

Similarly, for various real data sets Jolliffe (1973) found that none of the variable selection methods was uniformly best, but several of them, including B2 and B4, found reasonable subsets in most cases.

McCabe (1984) adopted a somewhat different approach to the variable selection problem. He started from the fact that, as has been seen in Chapters 2 and 3, PCs satisfy a number of different optimality criteria. A subset of the original variables which optimizes one of these criteria is termed a set of *principal variables* by McCabe (1984). Property A1 of Sections 2.1, 3.1, is uninteresting since it simply leads to a subset of variables whose variances are largest, but other properties lead to one of four criteria

(a) $$\text{Minimize } \prod_{j=1}^{m^*} \theta_j;$$

(b) $$\text{Minimize } \sum_{j=1}^{m^*} \theta_j;$$

(6.3.1)

(c) $$\text{Minimize } \sum_{j=1}^{m^*} \theta_j^2;$$

(d) $$\text{Maximize } \sum_{j=1}^{m^-} \rho_j^2;$$

where $\theta_j, j = 1, 2, \ldots, m^*$ are the eigenvalues of the conditional covariance (or correlation) matrix of the m^* deleted variables, given the values of the m selected variables, and $\rho_j, j = 1, 2, \ldots, m^- = \min(m, m^*)$ are the canonical correlations between the set of m^* deleted variables and the set of m selected variables.

Consider, for example, Property A4 of Sections 2.1 and 3.1, where $\det(\Sigma_y)$ (or $\det(S_y)$ for samples) is to be maximized. In PCA, y consists of ortho-normal linear functions of x; for principal variables y is a subset of x.

But, from a well-known result concerning partitioned matrices, $\det(\Sigma) = \det(\Sigma_y) \det(\Sigma_{y^*y})$, where Σ_{y^*y} is the matrix of conditional covariances for those variables not in $\mathbf{y}$, given the value of $\mathbf{y}$. Because Σ, and hence $\det(\Sigma)$, is fixed for a given random vector $\mathbf{x}$, maximizing $\det(\Sigma_y)$ is equivalent to minimizing $\det(\Sigma_{y^*y})$. Now $\det(\Sigma_{y^*y}) = \prod_{j=1}^{m^*} \theta_j$, so that Property A4 becomes criterion (6.3.1a) when deriving principal variables. Other properties of Chapters 2 and 3 can similarly be shown to be equivalent to one of the four criteria of (6.3.1), when dealing with principal variables.

Of the four criteria, McCabe (1984) argues that only for the first is it computationally feasible to explore all possible subsets, although the second can be used to define a stepwise variable-selection procedure; the other two criteria are not explored further in his paper.

Finally in this section, two further possible methods for variable selection are briefly discussed. Neither method uses PCs directly to select variables, so they are not described in detail here. However, both methods are related to topics discussed more fully in other chapters.

The RV-coefficient, due to Robert and Escoufier (1976), was discussed in Section 3.2. To use the coefficient to select a subset of variables, Robert and Escoufier (1976) suggest finding $\mathbf{X}_1$, which maximizes $RV(\mathbf{X}, \mathbf{M}'\mathbf{X}_1)$ where $RV(\mathbf{X}, \mathbf{Y})$ is defined by equation (3.2.2) of Section 3.2. $\mathbf{X}_1$ is the $(n \times m)$ submatrix of $\mathbf{X}$ consisting of n observations on a subset of m variables, and $\mathbf{M}$ is a specific $(m \times m)$ orthogonal matrix, whose construction is described in Robert and Escoufier's paper. No examples are given in the paper of how the method works in practice.

Hawkins and Eplett (1982) describe a method which can be used for selecting a subset of variables in regression; their technique, and an earlier one introduced by Hawkins (1973), is discussed in Sections 8.4 and 8.5. Hawkins and Eplett (1982) note that their method is also potentially useful for selecting a subset of variables in situations other than multiple regression, but, as with the RV-coefficient, no numerical example is given in the original paper.

6.4. Examples Illustrating Variable Selection

Two examples will be presented here; two other relevant examples are given in Section 8.7.

6.4.1. Alate adelges (Winged Aphids)

These data were first presented by Jeffers (1967) and comprise 19 different variables measured on 40 winged aphids. A description of the variables, together with correlation matrix and the coefficients of the first four PCs based on the correlation matrix, is given by Jeffers (1967) and will not be

Table 6.4. Subsets of selected variables, *Alate adelges.*
(Each row corresponds to a selected subset with × denoting a selected variable.)

		Variables						
	5	8	9	11	13	14	17	19
McCabe, using criterion (6.3.1a)								
Three variables { best				×			×	×
second best			×	×			×	
Four variables { best				×	×		×	×
second best	×			×	×			×
Jolliffe, using criteria B2, B4								
Three variables { B2			×		×		×	
B4				×	×	×		
Four variables { B2	×	×		×		×		
B4	×			×	×	×		

reproduced here. For 17 of the 19 variables, all of the correlation coefficients are positive, reflecting the fact that 12 variables are lengths or breadths of parts of each individual, and some of the other (discrete) variables also measure aspects of the size of each aphid. Not surprisingly, the first PC based on the correlation matrix accounts for a large proportion (73.0%) of the total variation, and this PC is a measure of overall size of each aphid. The second PC, accounting for 12.5% of total variation, has its largest coefficients on five of the seven discrete variables, and the third PC (3.9%) is almost completely dominated by one variable, number of antennal spines. This variable, which is one of the two variables which are negatively correlated with size, has a coefficient, in the third PC, five times as large as any other variable.

Table 6.4 gives various subsets of variables selected by Jolliffe (1973) and by McCabe (1982) in an earlier version of his 1984 paper, which included additional examples. The subsets given by McCabe (1982) are the best two according to criterion (6.3.1a), whereas those from Jolliffe (1973) are selected by the criteria B2 and B4 discussed above. Only the results for $m = 3$ are given in Jolliffe (1973) but Table 6.4 also gives results for $m = 4$ using his methods.

There is considerable overlap between the various subsets selected. In particular, variable 11 is a universal choice and variables 5 and 17 also appear in subsets selected by both authors' methods. Conversely, variables $\{1-4, 6, 7, 10, 12, 15, 16, 18\}$ appear in none of subsets of Table 6.4. It should be noted the variable 11 is 'number of antennal spines', which, as discussed above, dominates the third PC. Variables 5 and 17, measuring number of spiracles and number of ovipositor spines, respectively, are both among the group of variables on the second PC.

Regarding the choice of m, the l_k criterion of Section 6.1.2, was found by Jolliffe (1972), using simulation studies, to be appropriate for methods B2 and B4 with a cut-off close to $l^* = 0.7$. In the present examples the criterion suggests $m = 3$, since $l_3 = 0.75$ and $l_4 = 0.50$. Confirmation that m should be this small is given by the criterion t_k of Section 6.1.1. Two PCs account for 85.4% of the variation, three PCs give 89.4% and four PCs contribute 92.0%, from which Jeffers (1967) concludes that two PCs are sufficient to account for most of the variation. However, Jolliffe (1973) also looked at how well other aspects of the structure of data were reproduced for various values of m. For example, the form of the PCs, and the division into four distinct groups of aphids (see Section 9.2 for further discussion of this aspect) were both examined and found to be noticeably better reproduced for $m = 4$, than for $m = 2$ or 3, so it seems that the criteria of Sections 6.1.1 and 6.1.2 might be relaxed somewhat when very small values of m are indicated, especially when coupled with small values of n, the sample size. McCabe (1982) notes that four or five of the original variables are necessary in order to account for as much variation as the first two PCs, confirming that $m = 4$ or 5 is probably appropriate here.

6.4.2. Crime Rates

These data were given by Ahamad (1967), and consist of measurements of the crime rate in England and Wales for 18 different categories of crime (the variables) for the 14 years, 1950–63. The sample size $n = 14$ is very small, and in fact smaller than the number of variables. Furthermore, the data are time series, and the 14 observations are not independent (see Section 11.2), so that the effective sample size is even smaller than 14. Leaving aside this potential problem, and other criticisms of Ahamad's analysis (Walker, 1967), subsets of variables which were selected by McCabe (1982) and Jolliffe (1973) using the correlation matrix are shown in Table 6.5, which has the same format as Table 6.4.

Table 6.5. Subsets of selected variables, crime rates.
(Each row corresponds to a selected subset with × denoting a selected variable.)

	Variables								
	1	3	5	7	10	13	14	16	17
McCabe using criterion (6.3.1a)									
Three variables { best	×							×	×
{ second best	×						×		×
Four variables { best	×					×	×		×
{ second best	×				×	×	×		
Jolliffe using criteria B2, B4									
Three variables { B2	×			×		×			
{ B4	×	×	×						
Four variables { B2	×			×	×	×			
{ B4	×	×	×						×

There is a strong similarity between the correlation structure of the present data set and that of the previous example. Most of the variables considered increased during the time period considered, and the correlations between these variables are large and positive. (Some elements of the correlation matrix given by Ahamad (1967) are incorrect; Jolliffe (1970) gives the correct values.)

The first PC, based on the correlation matrix, therefore has large coefficients on all these variables, so it measures an 'average crime rate' based largely on 13 of the 18 variables, and it accounts for 71.7% of the total variation. The second PC, accounting for 16.1% of the total variation, has large coefficients on the five variables whose behaviour over the 14 years was 'atypical' in one way or another. The third PC, accounting for 5.5% of the total variation, is dominated by the single variable 'homicide', which stayed almost constant (compared with the trends in other variables) over the period of study. On the basis of t_k, only two or three PCs are necessary, since they account for 87.8%, 93.3%, respectively, of the total variation. The third and fourth eigenvalues are 0.96, 0.68, so that a cut-off of $l^* = 0.70$ will give $m = 3$, but l_4 is so close to 0.70 that caution suggests $m = 4$. Such conservatism is particularly appropriate for small sample sizes, as in this example, where sampling variation may be substantial. As in the previous example, Jolliffe (1973) finds that the inclusion of a fourth variable produces a marked improvement in reproducing some of the results given by all 18 variables. McCabe (1982) also indicates that $m = 3$ or 4 is appropriate.

The subsets chosen by McCabe's and Jolliffe's methods overlap less than in the previous example, and McCabe's subsets change noticeably in going from $m = 3$ to $m = 4$. However, there is still substantial agreement; for example, variable 1 is a member of all the selected subsets, and variables 10, 13, and 17 are selected by both types of method, whereas variables {2, 4, 6, 8, 9, 11, 12, 15, 18} are not selected at all.

Among the variables which are selected by more than one method, variable 1 is 'homicide', which dominates the third PC and is the only crime whose occurrence shows no evidence of serial correlation during the period 1950–63. Because its behaviour is different from that of all the other variables, it is important that it should be retained in any subset which seeks to account for most of the variation in **x**. Variables 10, 13 and 17 are again atypical of the general upward trend: variable 13 (assault) actually decreases between 1950 and 1963, variable 17 (indecent exposure) decreases for most years up to 1957, but increases in all but one year thereafter, and variable 10 (malicious injuries to property) increases in every year except one, but the 'one' is truly exceptional since there is a decrease of nearly 60%! This behaviour strongly suggests that variable 10 was redefined that year, and should not have been included in the analyses. However, given that it *was* included, its behaviour is certainly different from all the other variables, and it is therefore a strong candidate for inclusion in subsets of the variables.

In addition to the examples given here, Jolliffe (1973), McCabe (1984) and, especially, McCabe (1982) give further illustrations of variable selection based on PCs. Tortora (1980) presents an example which illustrates the effect on the non-iterative version of rule B2 of using unweighted or weighted data from a disproportionate stratified survey design.

CHAPTER 7

Principal Component Analysis and Factor Analysis

Principal component analysis has often been dealt with in textbooks as a special case of factor analysis, and this tendency has been continued by many computer packages which treat PCA as one option in a program for factor analysis—see Appendix A2. This view is misguided since PCA and factor analysis, as usually defined, are really quite distinct techniques. The confusion may have arisen, in part, because of Hotelling's (1933) original paper, in which principal components were introduced in the context of providing a small number of 'more fundamental' variables which determine the values of the p original variables. This is very much in the spirit of the factor model introduced in Section 7.1, although Girschick (1936) indicates that there were soon criticisms of Hotelling's method of PCs, as being inappropriate for factor analysis. Further confusion results from the fact that practitioners of 'factor analysis' do not always have the same definition of the technique (see Jackson, 1981). The definition adopted in this chapter is, however, fairly standard.

Both PCA and factor analysis aim to reduce the dimensionality of a set of data, but the approaches used to do so are different for the two techniques. Principal component analysis has been extensively used as part of factor analysis, but this involves 'bending the rules' which govern factor analysis, and there is much confusion in the literature over the similarities and differences between the techniques. This chapter will attempt to clarify the issues involved, and starts, in Section 7.1, with a definition of the basic model for factor analysis. Section 7.2 then discusses how a factor model may be estimated, and how PCs are, but should perhaps not be, used in this estimation process. Section 7.3 contains further discussion of differences between PCA and factor analysis, and Section 7.4 gives a numerical example, which compares the results of PCA and factor analysis. Finally, in Section 7.5, a few concluding remarks are made regarding the 'relative merits' of PCA and factor analysis, and the possible use of rotation with PCA.

7.1. Models for Factor Analysis

The basic idea underlying factor analysis is that p observed random variables, $\mathbf{x}$, can be expressed, except for an error term, as linear functions of m ($< p$) hypothetical (random) variables or *common factors*, i.e. if $x_1, x_2, \ldots, x_p$ are the variables and $f_1, f_2, \ldots, f_m$ are the factors then

$$
\begin{aligned}
x_1 &= \lambda_{11} f_1 + \lambda_{12} f_2 + \cdots + \lambda_{1m} f_m + e_1, \\
x_2 &= \lambda_{21} f_1 + \lambda_{22} f_2 + \cdots + \lambda_{2m} f_m + e_2, \\
&\vdots \\
x_p &= \lambda_{p1} f_1 + \lambda_{p2} f_2 + \cdots + \lambda_{pm} f_m + e_p,
\end{aligned}
\tag{7.1.1}
$$

where λ_{jk}, $j = 1, 2, \ldots, p$; $k = 1, 2, \ldots, m$ are constants called the *factor loadings*, and e_j, $j = 1, 2, \ldots, p$ are error terms, sometimes called *specific factors* (because e_j is 'specific' to x_j, whereas the f_k's are 'common' to several x_j's). Equation (7.1.1) can be rewritten in matrix form, with obvious notation, as

$$
\mathbf{x} = \Lambda \mathbf{f} + \mathbf{e}.
\tag{7.1.2}
$$

One contrast between PCA and factor analysis is immediately apparent. Factor analysis attempts to achieve a reduction from p to m dimensions by postulating a *model* relating $x_1, x_2, \ldots, x_p$ to m hypothetical variables. There is *no such explicit model* underlying PCA, even though Mandel (1972) argues that using PCA implies an implicit model—see Section 3.7.

The form of the basic model for factor analysis given in (7.1.2) is fairly standard, although some authors give somewhat different versions—for example, there could be *three* terms on the right-hand side corresponding to contributions from common factors, specific factors *and* measurement errors (Jöreskog et al. 1976, p. 57), or the model could be made non-linear. There are a number of assumptions associated with the factor models, as follows:

(i) $E[\mathbf{e}] = \mathbf{0}, \qquad E[\mathbf{f}] = \mathbf{0}, \qquad E[\mathbf{x}] = \mathbf{0}.$

Of these three assumptions, the first is a standard assumption for error terms in most statistical models, and the second is convenient and loses no generality. The third may not be true, but if it is not, (7.1.2) can be simply adapted to become

$$
\mathbf{x} = \boldsymbol{\mu} + \Lambda \mathbf{f} + \mathbf{e},
\tag{7.1.3}
$$

where $E[\mathbf{x}] = \boldsymbol{\mu}$. Equation (7.1.3) introduces only a slight amount of algebraic complication, compared with (7.1.2), but (7.1.2) loses no real generality, and is usually adopted.

(ii) $$E[\mathbf{ee}'] = \mathbf{\Psi} \quad \text{(diagonal)},$$

$$E[\mathbf{fe}'] = \mathbf{0} \quad \text{(a matrix of zeros)},$$

$$E[\mathbf{ff}'] = \mathbf{I}_m \quad \text{(an identity matrix)}.$$

The first of these three assumptions is merely stating that the error terms are uncorrelated, which is a basic assumption of the factor model i.e. all of $\mathbf{x}$ which is attributable to common influences is contained in $\mathbf{\Lambda f}$, and e_j, e_k, $j \neq k$ are therefore uncorrelated. The second assumption, that the common factors are uncorrelated with the specific factors, is also a fundamental one. However, the third assumption can be relaxed, so that the common factors may be correlated (oblique) rather than uncorrelated (orthogonal). Many techniques in factor analysis have been developed for finding orthogonal factors, but some authors, such as Cattell (1978, p. 128) argue that oblique factors are almost always necessary in order to get a correct factor structure. Such details will not be explored here, since the present objective is to compare factor analysis with PCA, rather than give a full description of factor analysis, and for convenience all three assumptions will be made.

(iii) For some purposes, such as testing for an appropriate value of m, it is necessary to make distributional assumptions. Usually the assumption of multivariate normality is made in such cases but, like PCA, many of the results of factor analysis do not depend on specific distributional assumptions.

(iv) Some restrictions are generally necessary on $\mathbf{\Lambda}$, because without any restrictions there will be a multiplicity of possible $\mathbf{\Lambda}$'s which give equally good solutions. This problem will be discussed further in the next section.

7.2. Estimation of the Factor Model

At first sight, the factor model (7.1.2) looks like a standard regression model such as that given in Property A6 of Section 3.1—see also Chapter 8. However, closer inspection reveals a substantial difference from the standard regression framework, namely that neither $\mathbf{\Lambda}$ nor $\mathbf{f}$ in (7.1.2) is known, whereas in regression $\mathbf{\Lambda}$ would be known, and $\mathbf{f}$ would contain the only unknown parameters. This means that different estimation techniques must be used, and it also means that there is indeterminacy in the solutions—the 'best-fitting' solution will not be unique.

Estimation of the model is usually done initially in terms of the parameters in $\mathbf{\Lambda}$ (and $\mathbf{\Psi}$), whereas estimates of $\mathbf{f}$ are found at a later stage. Given the assumptions of the previous section, the covariance matrix can be calculated for both sides of (7.1.2) giving

$$\mathbf{\Sigma} = \mathbf{\Lambda\Lambda}' + \mathbf{\Psi}. \qquad (7.2.1)$$

In practice, we have the sample covariance, or correlation, matrix, $\mathbf{S}$, rather than $\mathbf{\Sigma}$, and $\mathbf{\Lambda}$ and $\mathbf{\Psi}$ are found so as to satisfy

$$\mathbf{S} = \mathbf{\Lambda}\mathbf{\Lambda}' + \mathbf{\Psi}$$

(which does not involve the unknown vector of factor scores, $\mathbf{f}$), as closely as possible. The indeterminacy of the solution now becomes obvious, since if $\mathbf{\Lambda}$, $\mathbf{\Psi}$ is a solution of (7.2.1) and $\mathbf{T}$ is an orthogonal matrix, then $\mathbf{\Lambda}^*$, $\mathbf{\Psi}$ is also a solution, where $\mathbf{\Lambda}^* = \mathbf{\Lambda}\mathbf{T}$. This follows since

$$\mathbf{\Lambda}^*\mathbf{\Lambda}^{*'} = (\mathbf{\Lambda}\mathbf{T})(\mathbf{\Lambda}\mathbf{T})'$$

$$= \mathbf{\Lambda}\mathbf{T}\mathbf{T}'\mathbf{\Lambda}'$$

$$= \mathbf{\Lambda}\mathbf{\Lambda}', \quad \text{since } \mathbf{T} \text{ is orthogonal.}$$

Because of the indeterminacy, estimation of $\mathbf{\Lambda}$ and $\mathbf{\Psi}$ typically proceeds in two stages. In the first, some restrictions are placed on $\mathbf{\Lambda}$ in order to find a unique initial solution. Having found an initial solution, other solutions which can be found by *rotation* of $\mathbf{\Lambda}$ (i.e. multiplication by an orthogonal matrix $\mathbf{T}$) are explored, and the 'best' one chosen according to some particular criterion. There are several possible criteria, but all are designed to make the structure of $\mathbf{\Lambda}$ as simple as possible in some sense, with most elements of $\mathbf{\Lambda}$ either 'close to zero' or 'far from zero', with as few as possible of the elements taking intermediate values. Most large statistical computer packages provide options for several different rotation criteria, such as varimax, quartimax and promax. Cattell (1978, p. 136) gives a non-exhaustive list of eleven automatic rotation methods which includes some like oblimax, which enable the factors to become oblique, by allowing $\mathbf{T}$ to be not necessarily orthogonal. No details of particular methods of rotation will be presented here—see Cattell (1978, p. 136), Lawley and Maxwell (1971, Chapter 6), or Rummel (1970, Chapters 16 and 17)—but an example illustrating the results of two such methods is given in Section 7.4. It is also pointed out in Section 7.5 that, although 'standard' PCA does not include rotation, there may be circumstances in which rotation of a subset of the PCs can be advantageous.

The great advantage of rotation is that it simplifies the factor loadings or rotated PC coefficients, which can help in interpreting the factors or rotated PCs. This can be illustrated nicely using diagrams (see Figures 7.1 and 7.2) in the simple case where only $m = 2$ factors (PCs) are retained. Figure 7.1 plots the loadings of ten variables on two factors. In fact, these loadings are the coefficients $\mathbf{a}_1$, $\mathbf{a}_2$ for the first two PCs, normalized so that $\mathbf{a}_k'\mathbf{a}_k = l_k$ (where l_k is the kth eigenvalue of $\mathbf{S}$) rather than $\mathbf{a}_k'\mathbf{a}_k = 1$, for the example presented in detail later in the chapter. When an orthogonal rotation method (varimax) is performed, the loadings for the rotated factors are given by the projections of each plotted point onto the axes represented by dashes on Figure 7.1. Similarly, rotation using an oblique rotation method (direct quartimin) gives loadings after rotation by projecting onto the new axes shown on Figure 7.2.

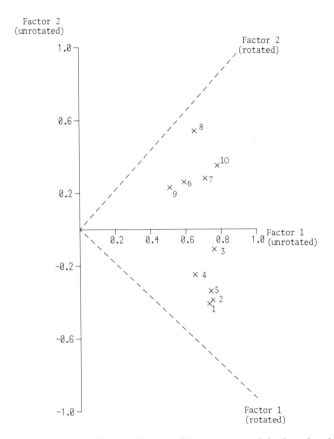

Figure 7.1. Factor loadings for two factors with respect to original, and orthogonally rotated, factors.

It is seen that in Figure 7.2 all points lie close to one or other of the axes, and so have near-zero loadings on the factor represented by the other axis, giving a very simple structure for the loadings. The loadings implied for the rotated factors in Figure 7.1, whilst having simpler structure than the original coefficients, are not so simple as those for Figure 7.2, thus illustrating the advantage of oblique, compared to orthogonal, rotation.

Returning to the first stage in the estimation of Λ and Ψ, there are a number of ways of constructing initial estimates. Some such as the centroid method (see Cattell, 1978, Section 2.3) were developed before the advent of computers, and were designed to give quick, computationally feasible, results. Such methods do a reasonable job of getting a crude factor model, but have little or no firm mathematical basis for doing so. This, among other aspects of factor analysis, gave it a 'bad name' among mathematicians and statisticians —Chatfield and Collins (1980, Chapter 5), for example, treat the topic rather

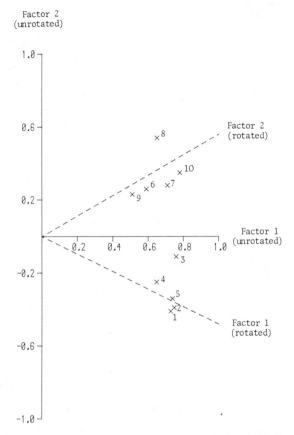

Figure 7.2. Factor loadings for two factors with respect to original, and obliquely rotated, factors.

dismissively, ending with the recommendation that factor analysis 'should not be used in most practical situations'.

There are more 'statistically respectable' approaches, such as the Bayesian approach outlined by Press (1972, Section 10.6.2) and the widely implemented idea of maximum likelihood estimation of Ψ and Λ, assuming multivariate normality of $\mathbf{f}$ and $\mathbf{e}$, as described in Lawley and Maxwell (1971, Chapter 4). Finding maximum likelihood estimates of Ψ and Λ leads to an iterative procedure, involving a moderate amount of algebra, which will not be repeated here—see, for example, Lawley and Maxwell (1971).

An interesting point is that factor loadings found by maximum likelihood for a correlation matrix are equivalent to those for corresponding covariance matrices, i.e. they are scale invariant. This is in complete contrast to what happens for PCA—see Sections 2.3 and 3.3.

A potential problem with the maximum likelihood approach is that it relies on the assumption of multivariate normality, which may not be justified. However, it can be shown (Morrison, 1976, Section 9.8; Rao, 1955, which is also reproduced in Bryant and Atchley (1975)) that the maximum likelihood estimators (MLEs) also optimize two criteria which make no direct distributional assumptions.

If the factor model (7.1.2) holds exactly, then the partial correlations between the elements of $\mathbf{x}$, given the value of $\mathbf{f}$, are zero (see also Section 6.1.6), since $\mathbf{f}$ accounts for all the common variation in the elements of $\mathbf{x}$. To derive the criterion described by Morrison (1976), a sample estimate of the matrix of partial correlations is calculated. The determinant of this matrix will attain its maximum value of one when all its off-diagonal elements are zero, so that maximizing this determinant is one way of attempting to minimize the absolute values of the partial correlations. It is this maximization problem which leads to the MLEs, but here they appear regardless of whether or not multivariate normality holds.

The procedure suggested by Rao (1955) is based on canonical correlation analysis (see Section 9.3) between $\mathbf{x}$ and $\mathbf{f}$. He looks, successively, for pairs of linear functions $\{\mathbf{a}'_{k1}\mathbf{x}, \mathbf{a}'_{k2}\mathbf{f}\}$ which have maximum correlation subject to being uncorrelated with previous pairs. The factor loadings are then proportional to the elements of the $\mathbf{a}_{k2}$'s $k = 1, 2, \ldots, m$, which in turn leads to the same loadings as for the MLEs based on the assumption of multivariate normality (Rao, 1955). As with the criterion based on partial correlations, no distributional assumptions are necessary for Rao's canonical analysis.

In a way, the behaviour of the partial correlation and canonical correlation criteria parallels the phenomenon in regression, where the least squares criterion is valid regardless of the distribution of error terms, but if errors are normally distributed, then least squares estimators have the added attraction of maximizing the likelihood function.

An alternative, but popular, way of getting initial estimates for Λ is to use the first m PCs. If $\mathbf{z} = \mathbf{A}'\mathbf{x}$ is the vector consisting of all p PCs, with $\mathbf{A}$ defined to have $\boldsymbol{\alpha}_k$, the kth eigenvector of Σ, as its kth column, as in (2.1.1), then $\mathbf{x} = \mathbf{A}\mathbf{z}$ because of the orthogonality of $\mathbf{A}$. If $\mathbf{A}$ is partitioned into its first m and last $p - m$ columns, with a similar partitioning of the rows of $\mathbf{z}$, then

$$\mathbf{x} = (\mathbf{A}_m \mid \mathbf{A}^*_{p-m})\left(\frac{\mathbf{z}_m}{\mathbf{z}^*_{p-m}}\right)$$

$$= \mathbf{A}_m\mathbf{z}_m + \mathbf{A}^*_{p-m}\mathbf{z}^*_{p-m}$$

$$= \Lambda\mathbf{f} + \mathbf{e}, \tag{7.2.2}$$

where

$$\Lambda = \mathbf{A}_m, \quad \mathbf{f} = \mathbf{z}_m \quad \text{and} \quad \mathbf{e} = \mathbf{A}^*_{p-m}\mathbf{z}^*_{p-m}.$$

Equation (7.2.2) looks very much like the factor model (7.1.2) but it violates a basic assumption of the factor model, because the elements of $\mathbf{e}$ in (7.2.2) are not usually uncorrelated. Despite the apparently greater sophistication of using the sample version of $\mathbf{A}_m$ as an initial estimator, compared with crude techniques such as centroid estimates, its theoretical justification is really no stronger.

As well as the straightforward use of PCs to estimate $\mathbf{\Lambda}$, many varieties of factor analysis use modifications of this approach; this topic will be discussed further in the next section.

7.3. Comparisons and Contrasts Between Factor Analysis and Principal Component Analysis

As mentioned in Section 7.1, a major distinction between factor analysis and PCA is that there is a model underlying factor analysis but no such model in PCA. Section 7.2 concludes by describing the most common way in which PCs are used in factor analysis. Further connections and contrasts between the two techniques are discussed in the present section.

Both techniques can be thought of as trying to represent some aspect of the covariance matrix $\mathbf{\Sigma}$ (or correlation matrix) as well as possible, but PCA concentrates on the diagonal elements, whereas in factor analysis the interest is in the off-diagonal elements. To justify this statement, consider first PCA. The objective is to maximize $\sum_{k=1}^{m} \mathrm{var}(z_k)$ or, since $\sum_{k=1}^{p} \mathrm{var}(z_k) = \sum_{j=1}^{p} \mathrm{var}(x_j)$, to account for as much as possible of the sum of diagonal elements of $\mathbf{\Sigma}$. As discussed after Property A3 in Section 2.1, the first m PCs will, in addition, often do a good job of explaining the off-diagonal elements of $\mathbf{\Sigma}$, which means that PCs can frequently provide an adequate initial solution in a factor analysis. However, this is not the stated purpose of PCA and will not hold universally. Turning now to factor analysis, consider the factor model (7.1.2) and the corresponding equation (7.2.1) for $\mathbf{\Sigma}$. It is seen that, since $\mathbf{\Psi}$ is diagonal, the common factor term $\mathbf{\Lambda f}$ in (7.1.2) accounts *completely* for the *off*-diagonal elements of $\mathbf{\Sigma}$ in the perfect factor model, but there is no compulsion for the diagonal elements to be well explained by the common factors. The elements, ψ_j, $j = 1, 2, \ldots, p$, of $\mathbf{\Psi}$ will *all* be low if *all* of the variables have considerable common variation, but if a variable, x_j, is almost independent of all other variables then $\psi_j = \mathrm{var}(e_j)$ will be almost as large as $\mathrm{var}(x_j)$. Thus, factor analysis concentrates on explaining only the off-diagonal elements of $\mathbf{\Sigma}$ by a small number of factors, whereas, conversely, PCA concentrates on the diagonal elements of $\mathbf{\Sigma}$.

This leads to another difference between the two techniques, concerning the number of dimensions, m, which give an adequate representation of the p-dimensional variable $\mathbf{x}$. In PCA, if any individual variables are almost independent of all other variables, then there will be a PC corresponding to each such variable, and the PC will be almost equivalent to the corresponding variable. Such 'single variable' PCs are generally included if an adequate representation of x is required, as was discussed in Section 6.1.6. In contrast, a common factor in factor analysis must contribute to *at least two* of the variables, so it is not possible to have a 'single variable' common factor. Instead, such factors appear as specific factors (error terms) and do not contribute to the dimensionality of the model. Thus, for a given set of data, the number of factors required for an adequate factor model will be no larger, and may be strictly smaller, than the number of PCs required to account for most of the variation in the data. If PCs are used as initial factors, then the ideal choice of m will often be less than that determined by most of the rules of Section 6.1.

The fact that a factor model concentrates on accounting for the off-diagonal elements, but not the diagonal elements, of Σ leads to various modifications of the idea of using the first m PCs to obtain initial estimates of factor loadings. Since the covariance matrix of the common factors' contribution to x is $\Sigma - \Psi$, it seems reasonable to use 'PCs' calculated for $\Sigma - \Psi$, rather than Σ, to construct initial estimates, leading to so-called *principal factor analysis*. This will, of course, require estimates of Ψ, which can be found in various ways (see, for example, Rummel, 1970, Chapter 13), either once-and-for-all or iteratively, leading to many different factor estimates. In fact, many of the different varieties of factor analysis correspond simply to using different estimates of Ψ in this type of 'modified PC' procedure. None of these estimates has a much stronger claim to absolute validity than does the use of the PCs of Σ, although arguments have been put forward to justify various different estimates of Ψ.

Another difference between PCA and factor analysis (after rotation) is that changing m, the dimensionality of the model, can have much more drastic effects on factor analysis than it does on PCA. In PCA, if m is increased from m_1 to m_2, say, then an additional $(m_2 - m_1)$ PCs are included, but the original m_1 PCs are still present and unaffected. However, in factor analysis an increase from m_1 to m_2 will produce m_2 factors, none of which need bear any resemblance to the original m_1 factors, though, typically, some will be similar if a factor model is really appropriate.

A final difference between PCs and common factors is that the former can be calculated exactly from x, whereas the latter typically cannot. The PCs are exact linear functions of x and can be calculated from

$$z = A'x.$$

The factors, however, are not exact linear functions of x; instead x is defined as a linear function of f apart from an error term, and when the relationship is reversed it certainly does not lead to an exact relationship between f and x. Indeed, the fact that the expected value of x is a linear function of f need not imply that the expected value of f is a linear function of x (unless multivariate normal assumptions are made). Thus, the use of PCs as initial factors may be forcing the factors into an unnecessarily restrictive linear framework. Because of the non-exactness of the relationship between f and x, the values of f, the *factor scores*, must be *estimated*, and there are several possible ways of doing this—see, for example, Lawley and Maxwell (1971, Chapter 8) or Morrison (1976, Section 9.10). Most methods of estimating factor scores use linear functions of **x**, but as pointed out by Jöreskog *et al.* (1976, p. 143) this can only give approximations to the 'true' scores.

To summarize, there are many ways in which PCA and factor analysis differ from one another. Despite these differences, they both have the aim of reducing the dimensionality of a vector of random variables, and the use of PCs to find initial factor loadings, though having no firm justification in theory, will often not be misleading in practice. In the special case where the elements of Ψ are proportional to the diagonal elements of Σ, Gower (1966) shows that the configuration of points produced by factor analysis will be similar to that found by PCA. In principal factor analysis, the results will be equivalent to those of PCA if all (non-zero) elements of Ψ are identical, (Rao, 1955). More generally, the coefficients found from PCA, and the loadings found from (orthogonal) factor analysis will often be very similar, although this will not hold unless all the elements of Ψ are of approximately the same size (Rao, 1955). Even though PCA and factor analysis may give similar numerical results for many examples, it is important to be aware of the differences between the two techniques and preferably to avoid the use of PCs in factor analysis since other, more acceptable, methods are widely available.

7.4. An Example of Factor Analysis

The example which follows is fairly typical of the sort of data which are often subjected to a factor analysis. The data were originally discussed by Yule *et al.* (1969) and consist of scores for 150 children on ten subtests of the Wechsler Pre-School and Primary Scale of Intelligence (WPPSI)—there are thus 150 observations on ten variables. The WPPSI tests were designed to measure 'intelligence' of children aged $4\frac{1}{2}$–6 years, and the 150 children tested in Yule *et al.*'s study were a sample of children who entered school in the Isle of Wight in the autumn of 1967, and who were tested during their second term in school. Their average age at the time of testing was 5 years, 5 months. Similar data sets are analysed in Lawley and Maxwell (1971).

Table 7.1 gives the variances and the coefficients of the first four PCs, when the analysis is done on the correlation matrix. It is seen that the first four components explain nearly 76% of the total variation, and that the variance of the fourth PC is 0.71. The fifth PC, with a variance of 0.51, will be discarded by most of the rules described in Section 6.1 and, indeed, in factor analysis it would be more usual to obtain only three, or perhaps only two, factors in the present example. Figures 7.1, 7.2 earlier in the chapter showed

Table 7.1. Coefficients for the first four PCs: children's intelligence tests.

Component number		1	2	3	4
	1	0.34	−0.39	0.09	−0.08
	2	0.34	−0.37	−0.08	−0.23
	3	0.35	−0.10	0.05	0.03
	4	0.30	−0.24	−0.20	0.63
Variable	5	0.34	−0.32	0.19	−0.28
number	6	0.27	0.24	−0.57	0.30
	7	0.32	0.27	−0.27	−0.34
	8	0.30	0.51	0.19	−0.27
	9	0.23	0.22	0.69	0.43
	10	0.36	0.33	−0.03	0.02
Eigenvalue		4.77	1.13	0.96	0.71
Cumulative percentage of total variation		47.7	59.1	68.6	75.7

the effect of rotation in this example when only two PCs are retained—here, where four PCs are retained, it will not be possible to represent the effect of rotation in the same, diagrammatic, way.

All of the correlations between the ten variables are positive so the first PC has the familiar pattern of being almost an equally weighted 'average' of all ten variables. The second PC contrasts the first five variables with the final five—this is not unexpected since these two sets of variables are of different types, namely 'verbal' tests and 'performance' tests respectively. The third PC is mainly a contrast between variables 6 and 9, which interestingly are the only two 'new' tests in the WPSSI battery, and the fourth does not have a very straightforward interpretation.

Table 7.2 gives the factor loadings when the first four PCs are rotated using one of the orthogonal rotation methods (varimax), and one of the oblique methods (direct quartimin). It would be counterproductive to give more varieties of factor analysis for this single example, since the differences in detail would tend to obscure the general conclusions which are drawn below. Many further examples can be found in texts on factor analysis such as Cattell (1978), Jöreskog et al. (1976), Lawley and Maxwell (1971) and Rummel (1970).

In order to make comparisons between Table 7.1 and Table 7.2 straightforward, the sum of squares of the PC coefficients and factor loadings are normalized to be equal to unity for each factor. Typically, the output from computer packages which implement factor analysis will not have this normalization—see Appendix A2. The correlations between the oblique factors in Table 7.2 are given in Table 7.3 and it can be seen that there is a non-trivial amount of correlation between the factors given by the oblique

Table 7.2. Rotated factor loadings—four factors:
children's intelligence tests.

Factor number		1	2	3	4
		Varimax			
	1	0.48	0.09	0.17	0.14
	2	0.49	0.15	0.18	−0.03
	3	0.35	0.22	0.24	0.22
	4	0.26	−0.00	0.64	0.20
Variable	5	0.49	0.16	0.02	0.15
number	6	0.05	0.34	0.60	−0.09
	7	0.20	0.51	0.18	−0.07
	8	0.10	0.54	−0.02	0.32
	9	0.10	0.13	0.07	0.83
	10	0.17	0.46	0.28	0.26
		Direct quartimin			
	1	0.51	−0.05	0.05	0.05
	2	0.53	0.04	0.05	−0.14
	3	0.32	0.13	0.16	0.15
	4	0.17	−0.19	0.65	0.20
Variable	5	0.54	0.06	−0.13	0.05
number	6	−0.07	0.28	0.67	−0.12
	7	0.16	0.53	0.13	−0.17
	8	0.03	0.62	−0.09	0.26
	9	0.00	0.09	0.02	0.87
	10	0.08	0.45	0.24	0.21

method. Despite this, the structure of the factor loadings is very similar for
the two factor rotation methods. The first factor in both methods has its
highest loadings in variables 1, 2, 3 and 5, with the next highest loadings on
variables 4 and 7. In factors 2, 3, 4 there is the same degree of similarity in the
position of the highest loadings: for factor 2, the loadings for variables 7, 8,
10, are highest, with an intermediate value on variable 6, factor 3 has large
loadings on variables 4 and 6 and an intermediate value on variable 10, and
factor 4 is dominated by variable 9 with intermediate values on variables 8

Table 7.3. Correlations between four
direct quartimin factors: children's
intelligence tests.

		Factor number		
		1	2	3
Factor	2	0.349		
number	3	0.418	0.306	
	4	0.305	0.197	0.112

Table 7.4. Factor loadings—three factors, varimax rotation: children's intelligence tests.

		Factor number		
		1	2	3
Variable number	1	0.47	0.09	0.14
	2	0.47	0.17	0.05
	3	0.36	0.23	0.24
	4	0.37	0.23	0.00
	5	0.45	0.08	0.23
	6	0.12	0.55	−0.05
	7	0.17	0.48	0.17
	8	0.05	0.36	0.52
	9	0.13	−0.01	0.66
	10	0.18	0.43	0.36

and 10. The only notable difference between the results for the two methods is that obliqueness allows the second method to achieve slightly higher values on the highest loadings and correspondingly lower values on the low loadings, as indeed it is meant to.

By contrast, the differences between the loadings before and after rotation are more substantial. After rotation, the 'general factor', with similar size coefficients on all variables, disappears, as do most negative coefficients, and the structure of the loadings is simplified. Again, this is precisely what rotation is meant to achieve.

To illustrate what happens when different numbers of factors are retained, Table 7.4 gives factor loadings for three factors using varimax rotation. The loadings for direct quartimin are again very similar. Before rotation, changing the number of PCs simply adds or deletes PCs leaving the remaining PCs unchanged. After rotation, however, deletion or addition of factors will usually change most, or all, of the factor loadings. In the present example, deletion of the fourth unrotated factor has left the first rotated factor almost unchanged, except for a modest increase in the loading for variable 4. Factor 2 is also similar to factor 2 in the four-factor analysis, although the resemblance is somewhat less strong than for factor 1. In particular, variable 6 now has the largest loading in factor 2, whereas it previously had only the fourth largest loading. The third factor in the three-factor solution is in no way similar to factor 3 in the four-factor analysis. In fact, it is quite similar to the original factor 4, and the original factor 3 has disappeared, with its highest loadings, on variables 4 and 6, partially 'transferred' to factors 1 and 2 respectively.

The behaviour displayed in this example, when a factor is deleted is not untypical of what happens in factor analysis generally, although the 'mixing-up' and 'rearrangement' of factors can be much more extreme than in the present case.

7.5. Concluding Remarks

It should be clear from the discussion of this chapter that it does not really make sense to ask whether PCA is 'better than' factor analysis or vice versa, because they are not direct competitors. If a model such as (7.1.2) seems a reasonable assumption for a data set, then factor analysis, rather than PCA, is appropriate. If no such model can be assumed, then factor analysis should not really be used.

Despite their different formulations and objectives, it can be informative to look at the results of both techniques on the same data set, as in the example above. Each technique will give different insights into the data structure, with PCA concentrating on explaining the diagonal elements, and factor analysis the off-diagonal elements, of the covariance matrix, and both may be useful. Furthermore, one of the main ideas of factor analysis, that of rotation, can be 'borrowed' for PCA, without any implication that a factor model is being assumed. Once PCA has been used to find an m-dimensional subspace which contains most of the variation in the original p variables, it is possible to redefine, by rotation, the axes (or variables) which form a basis for this subspace. The rotated variables will together account for the same amount of variation as the first few PCs, but will no longer successively account for the maximum possible variation. This behaviour is illustrated by Tables 7.1 and 7.2; the four rotated PCs in Table 7.2 together account for 75.7% of the total variation, as did the unrotated PCs in Table 7.1. However, the percentages of total variation accounted for by individual factors (rotated PCs) are 27.4, 21.9, 14.2 and 12.1, compared with 47.7, 11.3, 9.6 and 7.1 for the unrotated PCs. The rotated PCs, when expressed in terms of the original variables, may be easier to interpret than the PCs, because their coefficients will typically have a simpler structure. In addition, rotated PCs offer advantages compared to unrotated PCs in some types of analysis based on PCs— see Sections 8.5 and 10.1.

CHAPTER 8

Principal Components in Regression Analysis

As illustrated in the other chapters of this book, research continues into a wide variety of methods of using PCA in analysing various types of data. However, in no area has this research been more active in recent years, than in investigating approaches to regression analysis which use PCs in some form or another.

In multiple regression, one of the major difficulties with the usual least squares estimators is the problem of multicollinearity, which occurs when there are near-constant linear functions of two or more of the predictor, or regressor, variables. A recent, readable, review of the multicollinearity problem is given by Gunst (1983). Multicollinearities are often, but not always, indicated by large correlations between subsets of the variables, and, if multicollinearities exist, then the variances of some of the estimated regression coefficients can become very large, leading to unstable and often misleading estimates of the regression equation. To overcome this problem, various approaches have been proposed. One possibility is to use only a subset of the predictor variables, where the subset is chosen so that it does not contain multicollinearities. Numerous subset selection methods are available (see, for example, Draper and Smith, 1981, Chapter 6; Hocking, 1976; Miller, 1984), and among the methods are some based on PCs. These methods will be dealt with later in the chapter (Section 8.5), but, first, some more widely known uses of PCA in regression are described.

These uses follow from a second class of approaches to overcoming the problem of multicollinearity, namely the use of biased regression estimators. This class includes ridge regression, shrinkage estimators, and also approaches based on PCA. The best-known such approach, generally known as PC regression, simply starts by using the PCs of the predictor variables in place of the predictor variables. Since the PCs are uncorrelated, there are no multicollinearities between them, and the regression calculations are also simplified. If all the PCs are included in the regression, then the resulting model is equivalent to that obtained by least squares, so the large variances caused by multicollinearities have not been reduced. However, if some of the PCs are deleted from the regression equation, estimators of the coefficients in the original regression equation are obtained which are usually biased, but which simultaneously greatly reduce any large variances for regression coefficient estimators which have been caused by multicollinearities. Principal component regression is introduced in Section 8.1, and strategies for deciding which PCs to delete from the regression equation are discussed in Section 8.2; some connections between PC regression and other forms of biased regression are described in Section 8.3.

Variations on the basic idea of PC regression have also been proposed. One such variation, described in Section 8.3, allows the possibility that a PC may be only 'partly deleted' from the regression equation. A rather different approach, known as latent root regression, finds the PCs of the predictor variables *together with* the dependent variable. These PCs can then be used to construct biased regression estimators which will differ to some extent from those derived from PC regression. Latent root regression in various forms, together with its properties, is discussed in Section 8.4.

Yet another way in which PCA can be used in regression is to detect outliers. However, because the detection of outliers is important in other areas, as well as regression, discussion of this topic will be postponed until Section 10.1.

A topic which is related to, but different from, regression analysis is that of functional and structural relationships. The idea is, like regression analysis, to explore relationships between variables, but, unlike regression, the predictor variables as well as the dependent variable may be subject to error. Principal component analysis can again be used in investigating functional and structural relationships, and this topic is discussed in Section 8.6.

Finally in this chapter, in Section 8.7, two detailed examples are given of the use of PCs in regression, illustrating many of the techniques discussed in earlier sections.

8.1. Principal Component Regression

Consider the standard regression model, as defined in equation (3.1.5), i.e.

$$\mathbf{y} = \mathbf{X}\boldsymbol{\beta} + \boldsymbol{\varepsilon}, \tag{8.1.1}$$

where $\mathbf{y}$ is a vector of n observations on the dependent variable, measured about their mean, $\mathbf{X}$ is a $(n \times p)$ matrix whose (i, j)th element is the value of the jth predictor (or regressor) variable for the ith observation, again measured about its mean, $\boldsymbol{\beta}$ is a vector of regression coefficients and $\boldsymbol{\varepsilon}$ is the vector of error terms; the elements of $\boldsymbol{\varepsilon}$ are independent, each with the same variance σ^2. It is convenient to present the model (8.1.1) in 'centred' form, with all variables measured about their means. Furthermore, it is conventional in much of the literature on PC regression to assume that the predictor variables have been standardized so that $\mathbf{X'X}$ is proportional to the correlation matrix for the predictor variables, and this convention will be followed in the present chapter. Similar derivations to those below are possible if the predictor variables are in uncentred or non-standardized form, or if an alternative standardization has been used, but to save space and repetition, these derivations will not be given. Nor will we discuss the controversy that surrounds the choice of whether or not to centre the variables in a regression analysis. The interested reader is referred to Belsley (1984) and the discussion which follows that paper.

The values of the PCs for each observation are given by

$$\mathbf{Z} = \mathbf{XA}, \qquad (8.1.2)$$

where the (i, k)th element of $\mathbf{Z}$ is the value (score) of the kth PC for the ith observation, and $\mathbf{A}$ is a $(p \times p)$ matrix whose kth column is the kth eigenvector of $\mathbf{X'X}$.

Because $\mathbf{A}$ is orthogonal, $\mathbf{X\beta}$ can be rewritten as $\mathbf{XAA'\beta} = \mathbf{Z\gamma}$, where $\gamma = \mathbf{A'\beta}$. Equation (8.1.1) can therefore be written as

$$\mathbf{y} = \mathbf{Z\gamma} + \boldsymbol{\varepsilon}, \qquad (8.1.3)$$

which has simply replaced the predictor variables by their PCs in the regression model. Principal component regression can be defined as the use of the model (8.1.3) or of the reduced model

$$\mathbf{y} = \mathbf{Z}_m \boldsymbol{\gamma}_m + \boldsymbol{\varepsilon}_m, \qquad (8.1.4)$$

where $\boldsymbol{\gamma}_m$ is a vector of m elements which are a subset of elements of γ, $\mathbf{Z}_m$ is a $(n \times m)$ matrix whose columns are the corresponding subset of columns of $\mathbf{Z}$, and $\boldsymbol{\varepsilon}_m$ is the appropriate error term. Using least squares to estimate γ in (8.1.3), and then finding an estimate for $\boldsymbol{\beta}$ from the equation

$$\hat{\boldsymbol{\beta}} = \mathbf{A}\hat{\boldsymbol{\gamma}} \qquad (8.1.5)$$

is equivalent to finding $\hat{\boldsymbol{\beta}}$ by applying least squares directly to (8.1.1).

The idea of using PCs rather than the original predictor variables is not new (Hotelling, 1957; Kendall, 1957), and it has a number of advantages. First, calculating $\hat{\boldsymbol{\gamma}}$ from (8.1.3) is more straightforward than finding $\hat{\boldsymbol{\beta}}$ from (8.1.1) since the columns of $\mathbf{Z}$ are orthogonal, so that $\hat{\boldsymbol{\gamma}}$ becomes

$$\hat{\boldsymbol{\gamma}} = (\mathbf{Z'Z})^{-1}\mathbf{Z'y}$$

$$= \mathbf{L}^{-2}\mathbf{Z'y}, \qquad (8.1.6)$$

where $\mathbf{L}$ is the diagonal matrix whose kth diagonal element is $l_k^{1/2}$, and l_k is defined here as the kth largest eigenvalue of $\mathbf{X'X}$, rather than $\mathbf{S}$. Furthermore, if the regression equation is calculated for PCs, rather than for the predictor variables, then the contributions of each transformed variable (PC) to the equation can be more easily interpreted than the contributions of the original variables, because of orthogonality. Thus, even when multicollinearity is not a problem, regression on the PCs, rather than the original predictor variables, may have advantages for computation and interpretation. However, it should be noted that although interpretation of the separate contributions of each transformed variable is improved by taking PCs, the interpretation of the regression equation itself may be hindered if the PCs have no clear meaning.

The main advantage of PC regression occurs when multicollinearities are present. In this case, by deleting a subset of the PCs, especially those with small variances, much more stable estimates of β can be obtained. To see this, substitute (8.1.6) into (8.1.5) to give

$$\hat{\beta} = \mathbf{A}(\mathbf{Z}'\mathbf{Z})^{-1}\mathbf{Z}'\mathbf{y} \tag{8.1.7}$$

$$= \mathbf{A}\mathbf{L}^{-2}\mathbf{Z}'\mathbf{y}$$

$$= \mathbf{A}\mathbf{L}^{-2}\mathbf{A}'\mathbf{X}'\mathbf{y}$$

$$= \sum_{k=1}^{p} l_k^{-1}\mathbf{a}_k\mathbf{a}_k'\mathbf{X}'\mathbf{y}, \tag{8.1.8}$$

where l_k is the kth diagonal element of $\mathbf{L}^2$ and $\mathbf{a}_k$ is the kth column of $\mathbf{A}$. Equation (8.1.8) can also be derived more directly from $\hat{\beta} = (\mathbf{X}'\mathbf{X})^{-1}\mathbf{X}'\mathbf{y}$, by using the spectral decomposition (see Property A3 of Sections 2.1 and 3.1) of the matrix $(\mathbf{X}'\mathbf{X})^{-1}$, which has eigenvectors $\mathbf{a}_k$ and eigenvalues l_k^{-1}, $k = 1, 2, \ldots, p$.

Making the usual assumption that the elements of $\mathbf{y}$ are uncorrelated, each with the same variance σ^2 (i.e. the variance–covariance matrix of $\mathbf{y}$ is $\sigma^2\mathbf{I}_n$), it is seen from (8.1.7) that the variance–covariance matrix of $\hat{\beta}$ is

$$\sigma^2\mathbf{A}(\mathbf{Z}'\mathbf{Z})^{-1}\mathbf{Z}'\mathbf{Z}(\mathbf{Z}'\mathbf{Z})^{-1}\mathbf{A}' = \sigma^2\mathbf{A}(\mathbf{Z}'\mathbf{Z})^{-1}\mathbf{A}'$$

$$= \sigma^2\mathbf{A}\mathbf{L}^{-2}\mathbf{A}'$$

$$= \sigma^2\sum_{k=1}^{p} l_k^{-1}\mathbf{a}_k\mathbf{a}_k'. \tag{8.1.9}$$

This expression gives insight into how multicollinearities produce large variances for the elements of $\hat{\beta}$. If a multicollinearity exists, then it appears as a PC with very small variance (see also Sections 3.4 and 10.1); in other words, the later PCs have very small values of l_k (the variance of the kth PC is $l_k/(n-1)$, in the present notation), and hence very large values of l_k^{-1}. Thus (8.1.9) shows that any predictor variable which has moderate or large coefficients in any of the PCs associated with very small eigenvalues will have a very large variance.

One way of reducing this effect is to delete the terms from (8.1.8) which correspond to very small l_k's, leading to an estimator

$$\tilde{\beta} = \sum_{k=1}^{m} l_k^{-1}\mathbf{a}_k\mathbf{a}_k'\mathbf{X}'\mathbf{y}, \tag{8.1.10}$$

where $l_{m+1}, l_{m+2}, \ldots, l_p$ are the very small eigenvalues. This is equivalent to setting the last $p - m$ elements of γ equal to zero.

Then the variance–covariance matrix, $V(\tilde{\beta})$, for $\tilde{\beta}$ is

$$\sigma^2 \sum_{j=1}^m l_j^{-1} \mathbf{a}_j \mathbf{a}_j' \mathbf{X}' \mathbf{X} \sum_{k=1}^m l_k^{-1} \mathbf{a}_k \mathbf{a}_k'.$$

Substituting

$$\mathbf{X}'\mathbf{X} = \sum_{i=1}^p l_i \mathbf{a}_i \mathbf{a}_i'$$

from the spectral decomposition of $\mathbf{X}'\mathbf{X}$, we have

$$V(\tilde{\beta}) = \sigma^2 \sum_{i=1}^p \sum_{j=1}^m \sum_{k=1}^m l_i l_j^{-1} l_k^{-1} \mathbf{a}_j \mathbf{a}_j' \mathbf{a}_i \mathbf{a}_i' \mathbf{a}_k \mathbf{a}_k'.$$

Because the vectors $\mathbf{a}_i$, $i = 1, 2, \ldots, p$ are orthonormal, the only non-zero terms in the triple summation occur when $i = j = k$, so that

$$V(\tilde{\beta}) = \sigma^2 \sum_{k=1}^m l_k^{-1} \mathbf{a}_k \mathbf{a}_k'. \qquad (8.1.11)$$

If none of the first m l_k's are very small, then none of the variances given by the diagonal elements of (8.1.11) will be large.

The decrease in variance for the estimator $\tilde{\beta}$ given by (8.1.10), compared with the variance of $\hat{\beta}$, is achieved at the expense of introducing bias into the estimator $\tilde{\beta}$. This follows since

$$\tilde{\beta} = \hat{\beta} - \sum_{k=m+1}^p l_k^{-1} \mathbf{a}_k \mathbf{a}_k' \mathbf{X}' \mathbf{y}, \qquad E(\hat{\beta}) = \beta,$$

and

$$E\left[\sum_{k=m+1}^p l_k^{-1} \mathbf{a}_k \mathbf{a}_k' \mathbf{X}' \mathbf{y} \right] = \sum_{k=m+1}^p l_k^{-1} \mathbf{a}_k \mathbf{a}_k' \mathbf{X}' \mathbf{X} \beta$$

$$= \sum_{k=m+1}^p \mathbf{a}_k \mathbf{a}_k' \beta.$$

This last term is, in general, non-zero, so that $E(\tilde{\beta}) \neq \beta$. However, if multi-collinearity is a serious problem, the reduction in variance can be substantial, whereas the bias introduced may be comparatively small. In fact, if the elements of γ corresponding to deleted components are actually zero, then no bias will be introduced.

As well as, or instead of, deleting terms from (8.1.8) corresponding to small eigenvalues, it is also possible to delete terms for which the corresponding element of γ is not significantly different from zero. The selection of which elements are significantly non-zero is essentially a variable selection problem, with PCs as variables rather than the original predictor variables, and any of the well-known methods of variable selection for regression (see, for example,

Draper and Smith, 1981, Chapter 6) can be used. However, the problem is complicated by the desirability of also deleting high-variance terms from (8.1.8).

The definition of PC regression given above in terms of equations (8.1.3) and (8.1.4), is equivalent to using the linear model (8.1.1) and estimating $\boldsymbol{\beta}$ by

$$\tilde{\boldsymbol{\beta}} = \sum_M l_k^{-1} \mathbf{a}_k \mathbf{a}_k' \mathbf{X}' \mathbf{y}, \qquad (8.1.12)$$

where M is some subset of the integers $1, 2, \ldots, p$. Equation (8.1.10) is a special case of (8.1.12) with $M = \{1, 2, \ldots, m\}$, but, in the general definition of PC regression, M can be any subset of the first p integers, so that any subset of the coefficients of $\boldsymbol{\gamma}$, corresponding to the complement of M, can be set to zero. The next section will consider various strategies for choosing M, but we first note that once again the singular value decomposition (SVD) of $\mathbf{X}$, defined in Section 3.5, can be a useful concept—see also Sections 5.3, 6.1.5, 11.4, 11.5, 11.6 and Appendix A1. In the present context, it can be used to provide an alternative formulation of equation (8.1.12), and to help in the interpretation of the results of a PC regression. Assuming that $n \geq p$, and that $\mathbf{X}$ has rank p, recall that the SVD writes $\mathbf{X}$ in the form

$$\mathbf{X} = \mathbf{ULA}',$$

where

(i) $\mathbf{A}$ and $\mathbf{L}$ are as defined earlier in this section;
(ii) the columns of $\mathbf{U}$ are those eigenvectors of $\mathbf{XX}'$ which correspond to non-zero eigenvalues, normalized so that $\mathbf{U}'\mathbf{U} = \mathbf{I}_p$.

Then $\mathbf{X\beta}$ can be rewritten $\mathbf{ULA}'\boldsymbol{\beta} = \mathbf{U\delta}$, where $\boldsymbol{\delta} = \mathbf{LA}'\boldsymbol{\beta}$, so that $\boldsymbol{\beta} = \mathbf{AL}^{-1}\boldsymbol{\delta}$. The least squares estimator for $\boldsymbol{\delta}$ is

$$\hat{\boldsymbol{\delta}} = (\mathbf{U}'\mathbf{U})^{-1}\mathbf{U}'\mathbf{y} = \mathbf{U}'\mathbf{y},$$

leading to $\hat{\boldsymbol{\beta}} = \mathbf{AL}^{-1}\hat{\boldsymbol{\delta}}$.

The relationship between $\boldsymbol{\gamma}$, defined earlier, and $\boldsymbol{\delta}$ is straightforward, namely

$$\boldsymbol{\gamma} = \mathbf{A}'\boldsymbol{\beta} = \mathbf{A}'(\mathbf{AL}^{-1}\boldsymbol{\delta}) = (\mathbf{A}'\mathbf{A})\mathbf{L}^{-1}\boldsymbol{\delta} = \mathbf{L}^{-1}\boldsymbol{\delta},$$

so that setting a subset of elements of $\boldsymbol{\delta}$ equal to zero is equivalent to setting the same subset of elements of $\boldsymbol{\gamma}$ equal to zero, leading to an alternative formulation for (8.1.12). This formulation means that the SVD can provide an alternative computational approach for estimating PC regression equations. This is an advantage, since efficient algorithms exist for finding the SVD of a matrix—see Appendix A1.

Interpretation of the results of a PC regression can also be aided by using the SVD, as illustrated by Mandel (1982) for artificial data—see also Nelder (1985).

8.2. Strategies for Selecting Components in Principal Component Regression

When choosing the subset M in equation (8.1.12) there are two partially conflicting objectives. In order to eliminate large variances due to multi-collinearities, it is essential to delete all those components whose variances are very small but, at the same time, it is undesirable to delete components which have large correlations with the dependent variable y. One strategy for choosing M is simply to delete all those components whose variances are less than l^*, where l^* is some cut-off level. The choice of l^* is rather arbitrary, but when dealing with correlation matrices, where the average value of the eigenvalues is 1, a value of l^* somewhere in the range 0.01 to 0.1 seems to be useful in practice.

An apparently more sophisticated way of choosing l^* is to look at so-called variance inflation factors (VIFs) for the p predictor variables. The VIF for the jth variable when using standardized variables is defined as c_{jj}/σ^2 (which equals the jth diagonal element of $(\mathbf{X}'\mathbf{X})^{-1}$—Marquardt, 1970), where c_{jj} is the variance of the jth element of the least squares estimator for $\boldsymbol{\beta}$. If all the variables are orthogonal, then all the VIFs are equal to 1 but, if severe multicollinearities exist, then the VIFs for $\hat{\boldsymbol{\beta}}$ will be very large for those variables involved in the multicollinearities. By successively deleting the last few terms in (8.1.8), the VIFs will be reduced; deleting could continue until all VIFs are below some desired level. The VIF for a variable is related to the squared multiple correlation R^2, between that variable and the other $(p - 1)$ predictor variables by the formula $\text{VIF} = (1 - R^2)^{-1}$. Values of $\text{VIF} > 10$ correspond to $R^2 > 0.90$, and $\text{VIF} > 4$ is equivalent to $R^2 > 0.75$, so that values of R^2 can be considered when choosing how small a level of VIF is desirable. However, the choice of this desirable level is almost as arbitrary as the choice of l^* above.

Deletion based solely on variance is an attractive and simple strategy, and Property A6 of Section 3.1 gives it, at first sight, an added respectability. However, low variance for a component does not necessarily imply that the corresponding component is unimportant in the regression model. For example, Kung and Sharif (1980) gave an example from meteorology where, in a regression of monsoon onset date on all of the (ten) PCs, the most important PCs for prediction are (in decreasing order of importance) the eighth, second and tenth—see Table 8.1. The tenth component accounts for less than 1% of the total variation in the predictor variables, but is an important predictor of the dependent variable, and the most important PC in the regression accounts for 24% of the variation in y, but only 3% of the variation in $\mathbf{x}$. Further examples of this type are listed in Jolliffe (1982). Thus, the two objectives of deleting PCs with small variances, and of retaining PCs which are good predictors of the dependent variable, may not be simultaneously achievable.

Table 8.1. Variation accounted for by PCs of predictor variables in monsoon data for: (a) predictor variables; (b) dependent variable.

Component number			1	2	3	4	5	6	7	8	9	10
Percentage variation accounted for	(a)	Predictor variables	26	22	17	11	10	7	4	3	1	<1
	(b)	Dependent variable	3	22	<1	1	3	3	6	24	5	20

Some authors (e.g. Hocking, 1976; Mosteller and Tukey, 1977, pp. 397–398; Gunst and Mason, 1980, pp. 327–328) argue that the choice of PCs in the regression should be made entirely, or mainly, on the basis of variance reduction but, as can be seen from the examples cited by Jolliffe (1982), such a procedure can be dangerous if low-variance components have predictive value. Jolliffe (1982) suggests that examples where this occurs seem to be not uncommon in practice. Berk's (1984) experience with six data sets indicates the opposite conclusion, but it should be noted that several of his data sets are of a special type, in which strong positive correlations exist between all the regressor variables, *and* between the dependent variable and the regressor variable. In such cases the first PC will be a (weighted) average of the regressor variables, with all weights positive (see Section 3.8), and since y is also positively correlated with each regressor variable, it will be strongly correlated with the first PC.

In contrast to selection based solely on size of variance, the opposite extreme is to base selection only on values of t-statistics measuring the (independent) contribution of each PC to the regression equation. A compromise would be to delete PCs sequentially starting with the smallest variance, then the next smallest variance and so on; deletion would stop when the first significant t-value was reached. Such a strategy is likely to retain more PCs than are really necessary.

Hill *et al.* (1977) give a comprehensive discussion of various, more sophisticated, strategies for deciding which PCs to delete from the regression equation. Their criteria are of two main types depending on whether the primary objective is to get $\tilde{\beta}$ close to β, or to get $X\tilde{\beta}$, the estimate of y, close to y, (or to $E(y)$). In the first case estimation of β is the main interest; in the second it is prediction of y which is the chief concern. Whether or not $\tilde{\beta}$ is an improvement on $\hat{\beta}$ is determined, for several of the criteria, by looking at mean square error (MSE), so that variance and bias are both taken into account.

More specifically, two criteria are suggested of the first type, the 'weak' and 'strong' criteria. The weak criterion, due to Wallace (1972), prefers $\tilde{\beta}$ to $\hat{\beta}$ if $\text{tr}[\text{MSE}(\tilde{\beta})] \leq \text{tr}[\text{MSE}(\hat{\beta})]$, where $\text{MSE}(\tilde{\beta})$ is the matrix $E[(\tilde{\beta} - \beta)(\tilde{\beta} - \beta)']$ with a similar definition for the matrix $\text{MSE}(\hat{\beta})$. This simply means that $\tilde{\beta}$ is preferred when the expected Euclidean distance between $\tilde{\beta}$ and β is smaller than that between $\hat{\beta}$ and β.

The strong criterion insists that

$$MSE(\mathbf{c}'\tilde{\boldsymbol{\beta}}) \leq MSE(\mathbf{c}'\hat{\boldsymbol{\beta}})$$

for every non-zero p-element vector $\mathbf{c}$, where

$$MSE(\mathbf{c}'\tilde{\boldsymbol{\beta}}) = E[(\mathbf{c}'\tilde{\boldsymbol{\beta}} - \mathbf{c}'\boldsymbol{\beta})^2],$$

with, again, a similar definition for $MSE(\mathbf{c}'\hat{\boldsymbol{\beta}})$.

Among those criteria of the second type (where prediction of $\mathbf{y}$ rather than estimation of $\boldsymbol{\beta}$ is the main concern) which are considered by Hill *et al.* (1977), there are again two which use MSE. The first is also due to Wallace (1972) and is again termed a 'weak' criterion. It prefers $\tilde{\boldsymbol{\beta}}$ to $\hat{\boldsymbol{\beta}}$ if

$$E[(\mathbf{X}\tilde{\boldsymbol{\beta}} - \mathbf{X}\boldsymbol{\beta})'(\mathbf{X}\tilde{\boldsymbol{\beta}} - \mathbf{X}\boldsymbol{\beta})] \leq E[(\mathbf{X}\hat{\boldsymbol{\beta}} - \mathbf{X}\boldsymbol{\beta})'(\mathbf{X}\hat{\boldsymbol{\beta}} - \mathbf{X}\boldsymbol{\beta})],$$

so that $\tilde{\boldsymbol{\beta}}$ is preferred to $\hat{\boldsymbol{\beta}}$ if the expected Euclidean distance between $\mathbf{X}\tilde{\boldsymbol{\beta}}$ (the estimate of $\mathbf{y}$) and $\mathbf{X}\boldsymbol{\beta}$ (the expected value of $\mathbf{y}$) is smaller than the corresponding distance between $\mathbf{X}\hat{\boldsymbol{\beta}}$ and $\mathbf{X}\boldsymbol{\beta}$. An alternative MSE criterion is to look at the distance between each estimate of $\mathbf{y}$ and the actual, rather than expected, value of $\mathbf{y}$. Thus $\tilde{\boldsymbol{\beta}}$ is preferred to $\hat{\boldsymbol{\beta}}$ if

$$E[(\mathbf{X}\tilde{\boldsymbol{\beta}} - \mathbf{y})'(\mathbf{X}\tilde{\boldsymbol{\beta}} - \mathbf{y})] \leq E[(\mathbf{X}\hat{\boldsymbol{\beta}} - \mathbf{y})'(\mathbf{X}\hat{\boldsymbol{\beta}} - \mathbf{y})].$$

Substituting $\mathbf{y} = \mathbf{X}\boldsymbol{\beta} + \boldsymbol{\varepsilon}$ it follows that

$$E[(\mathbf{X}\tilde{\boldsymbol{\beta}} - \mathbf{y})'(\mathbf{X}\tilde{\boldsymbol{\beta}} - \mathbf{y})] = E[(\mathbf{X}\tilde{\boldsymbol{\beta}} - \mathbf{X}\boldsymbol{\beta})'(\mathbf{X}\tilde{\boldsymbol{\beta}} - \mathbf{X}\boldsymbol{\beta})] + n\sigma^2,$$

with a similar expression for $\hat{\boldsymbol{\beta}}$. At first sight, it would seem that this second criterion is equivalent to the first. However σ^2 is unknown, and although it can be estimated, we may get different estimates when the equation is fitted using $\tilde{\boldsymbol{\beta}}$, $\hat{\boldsymbol{\beta}}$ respectively.

Hill *et al.* (1977) considered several other criteria; further details may be found in their paper, which also describes connections between the various decision rules for choosing M, and gives illustrative examples. It is argued by Hill *et al.* (1977) that the choice of PCs should not be based solely on the size of their variance, but little advice is given on which of their criteria gives an overall 'best' trade-off between variance and bias; rather, separate circumstances are identified in which each may be the most appropriate.

Yet another approach to deletion of PCs which takes into account both variance and bias is given by Lott (1973). This approach simply calculates the adjusted multiple coefficient of determination,

$$\bar{R}^2 = 1 - \frac{(n-1)}{(n-p-1)}(1 - R^2),$$

where R^2 is the usual multiple coefficient of determination (squared multiple correlation) for the regression equation obtained from each subset, M, of interest. The 'best' subset is then the one which maximizes $\bar{R}^2$. Lott (1973) demonstrated that this very simple procedure worked well in a limited simulation study.

It is difficult to give any general advice regarding the choice of a decision rule for determining M. It is clearly inadvisable to base the decision entirely on the size of variance; conversely, inclusion of highly predictive PCs can also be dangerous if they also have very small variances, because of the resulting instability of the estimated regression equation. Use of MSE criteria provide a number of compromise solutions, but they are essentially arbitrary.

What PC regression *can* do, which least squares cannot, is to indicate explicitly whether a problem exists with respect to the removal of multicollinearity, i.e. whether instability in the regression coefficients can only be removed by simultaneously losing a substantial proportion of the predictability of y. The choice of M for PC regression, however, remains an open question.

8.3. Some Connections Between Principal Component Regression and Other Biased Regression Methods

Using the expressions (8.1.8), (8.1.9) for $\hat{\boldsymbol{\beta}}$, and its variance–covariance matrix, it was seen, in the previous section, that deletion of the last few terms from the summation for $\hat{\boldsymbol{\beta}}$ could dramatically reduce the high variances of elements of $\hat{\boldsymbol{\beta}}$ caused by multicollinearities. However, if any of the elements of $\boldsymbol{\gamma}$ corresponding to deleted components are non-zero, then the PC estimator $\tilde{\boldsymbol{\beta}}$ for $\boldsymbol{\beta}$ is biased. Various other methods of biased estimation which aim to remove collinearity-induced high variances have also been proposed. A full description of these methods will not be given here since several do not involve PCs directly, but there are various relationships between PC regression and other biased regression methods which will be briefly discussed.

Consider first ridge regression, which was described by Hoerl and Kennard (1970a, b) and which has since been the subject of much debate in the statistical literature. The estimator of $\boldsymbol{\beta}$ using the technique can be written, among other ways, as

$$\hat{\boldsymbol{\beta}}_R = \sum_{k=1}^{p} (l_k + \kappa)^{-1} \mathbf{a}_k \mathbf{a}_k' \mathbf{X}' \mathbf{y},$$

where κ is some fixed positive constant and the other terms in the expression have the same meaning as in (8.1.8.). The variance–covariance matrix of $\hat{\boldsymbol{\beta}}_R$ is equal to

$$\sigma^2 \sum_{k=1}^{p} l_k (l_k + \kappa)^{-2} \mathbf{a}_k \mathbf{a}_k'.$$

Thus, ridge regression estimators have rather similar expressions to those for least squares and PC estimators, but variance reduction is achieved not by deleting components, but by giving less weight to the later components. A generalization of ridge regression has p constants κ_k, $k = 1, 2, \ldots, p$ which must be chosen, rather than a single constant κ.

A modification of PC regression, due to Marquardt (1970) uses a similar, but more restricted, idea. Here, a PC regression estimator of the form (8.1.10) is adapted so that M includes the first m integers, excludes the integers $m + 2$, $m + 3, \ldots, p$, but includes the term corresponding to integer $(m + 1)$ with a weighting less than unity. Such estimators, and detailed discussion of them, are given by Marquardt (1970).

Ridge regression estimators 'shrink' the least squares estimators towards the origin, and so are similar in effect to the shrinkage estimators proposed by Stein (1960) and Sclove (1968). These latter estimators start with the idea of shrinking some or all of the elements of $\hat{\gamma}$ (or $\hat{\beta}$), using arguments based on loss functions, admissibility and prior information; choice of shrinkage constants is based on optimization of MSE criteria.

All these various biased estimators have relationships between them. In particular, all the present estimators, as well as latent root regression which is discussed in the next section, can be viewed as optimizing $(\tilde{\beta} - \beta)'X'X(\tilde{\beta} - \beta)$, subject to different constraints for different estimators (see Hocking, 1976).

Hocking (1976) also demonstrates that if the data set is augmented by a set of dummy observations, and least squares is used to estimate β from the augmented data, then generalized ridge, ridge, PC regression, Marquardt's modification and shrinkage estimators will all appear as special cases, for particular choices of the dummy observations and their variances. In a slightly different approach to the same topic, Hocking *et al.* (1976) give a broad class of biased estimators, which includes all the above estimators, including those derived from PC regression, as special cases. Oman (1978) shows how several biased regression methods, including PC regression, can be fitted into a Bayesian framework by using different prior distributions for β; Leamer and Chamberlain (1976) also look at a Bayesian approach to regression, and its connections with PC regression. Further, different, biased estimators have been suggested and compared with PC regression by Iglarsh and Cheng (1980) and Trenkler (1980), and relationships between ridge regression and PC regression are explored further by Hsuan (1981).

Essentially the same problem arises for all these biased methods as occurred in the choice of M for PC regression, namely what compromise should be chosen, in the trade-off between bias and variance. In ridge regression, this compromise manifests itself in the choice of κ, and for shrinkage estimators the amount of shrinkage must be determined. Suggestions have been made regarding rules for making these choices, but the decision is again still somewhat arbitrary.

8.4. Variations on Principal Component Regression

Marquardt's (1970) fractional rank estimator, which was described in the previous section, is one modification of PC regression as defined in Section 8.1, but it is a fairly minor modification. A rather different type of approach, which nevertheless still uses PCs in a regression problem, is provided by latent root regression. The main difference between this technique, and straightforward PC regression, is that the PCs are not calculated for the set of p predictor variables alone. Instead, they are calculated for a set of $(p + 1)$ variables, consisting of the p predictor variables *and* the dependent variable. This idea was suggested independently by Hawkins (1973) and by Webster *et al.* (1974), and termed 'latent root regression' by the latter authors. Subsequent papers (Gunst *et al.*, 1976; Gunst and Mason, 1977a) have investigated the properties of latent root regression, and compared it with other biased regression estimators. As with the other biased estimators discussed in the previous section, the latent root regression estimator can be derived by optimizing a quadratic function of $\boldsymbol{\beta}$, subject to constraints (Hocking, 1976). Latent root regression, as defined in Gunst and Mason (1980, Section 10.2), will now be described; the technique introduced by Hawkins (1973) has slight differences and will be discussed later in this section.

In latent root regression, a PCA is done on the set of $(p + 1)$ variables, described above, and the PCs corresponding to the smallest eigenvalues are examined. Those for which the coefficient of the dependent variable, y, is also small are called *non-predictive multicollinearities*, and are deemed to be of no use in predicting y. However, any PC with a small eigenvalue *will* be of predictive value if its coefficient for y is large. Thus, latent root regression tries to delete PCs which indicate multicollinearities, but only if the multicollinearities are useless for predicting y.

Let $\boldsymbol{\delta}_k$ be the vector of the p coefficients on the p predictor variables in the kth PC for the enlarged set of $(p + 1)$ variables; let δ_{0k} be the corresponding coefficient of y, and let $\tilde{l}_k$ be the corresponding eigenvalue. Then the latent root estimator for $\boldsymbol{\beta}$ is defined as

$$\hat{\boldsymbol{\beta}}_{\text{LR}} = \sum_{M_{\text{LR}}} f_k \boldsymbol{\delta}_k, \tag{8.4.1}$$

where M_{LR} is the subset of the integers 1, 2, ..., $p + 1$, in which integers corresponding to the non-predictive multicollinearities defined above, and no others, are deleted; the f_k are coefficients chosen to minimize residual sum of squares among estimators of the form (8.4.1).

The f_k can be determined by first using the kth PC to express $\mathbf{y}$ as a linear function of $\mathbf{X}$, and using this as an estimator $\hat{\mathbf{y}}_k$. A weighted average, $\hat{\mathbf{y}}_{\text{LR}}$, of the $\hat{\mathbf{y}}_k$, for $k \in M_{\text{LR}}$, is then constructed, where the weights are chosen so as to minimize the residual sum of squares $(\hat{\mathbf{y}}_{\text{LR}} - \mathbf{y})'(\hat{\mathbf{y}}_{\text{LR}} - \mathbf{y})$. $\hat{\mathbf{y}}_{\text{LR}}$ is then equal to the latent root regression estimator $\mathbf{X}\hat{\boldsymbol{\beta}}_{\text{LR}}$ and the f_k are given by

$$f_k = -\delta_{0k} \eta_y \tilde{l}_k^{-1} \left(\sum_{M_{\text{LR}}} \delta_{0k}^2 \tilde{l}_k^{-1} \right)^{-1}, \tag{8.4.2}$$

where $\eta_y^2 = \sum_{i=1}^{n} (y_i - \bar{y})^2$, and δ_{0k}, $\tilde{l}_k$ are as defined above. Note that the least squares estimator, $\hat{\boldsymbol{\beta}}$, can also be written in the form (8.4.1), if M_{LR} in (8.4.1) and (8.4.2) is taken to be the full set of PCs.

The full derivation of this expression for f_k is fairly lengthy, and can be found in Webster *et al.* (1974). It is interesting to note that f_k is proportional to the size of the coefficient of y in the kth PC, and inversely proportional to the variance of the kth PC, both of which relationships are intuitively reasonable.

In order to choose the subset M_{LR}, it is necessary to decide not only how small the eigenvalues must be in order to indicate multicollinearities, but also how large the coefficient of y must be in order to indicate a *predictive* multi-collinearity. Again, these are arbitrary choices, and *ad hoc* rules have been used, for example, by Gunst *et al.* (1976). A more formal procedure for identifying non-predictive multicollinearities is described by White and Gunst (1979), but its derivation is based on *asymptotic* properties of the statistics used in latent root regression.

Gunst *et al.* (1976) compared $\hat{\beta}_{LR}$ and $\hat{\beta}$ in terms of MSE, using a simulation study, for cases where there is only one multicollinearity, and found that $\hat{\beta}_{LR}$ showed substantial improvement over $\hat{\beta}$ when the multicollinearity is non-predictive. However, in cases where the single multicollinearity had some predictive value, the results were, unsurprisingly, less favourable to $\hat{\beta}_{LR}$. Gunst and Mason (1977a) reported a larger simulation study which compared PC, latent root, ridge and shrinkage estimators, again on the basis of MSE. Overall, latent root estimators again did well in many, but not all, situations studied, as did PC estimators, but no simulation study can ever be exhaustive, and different conclusions might possibly be drawn for other types of simulated data.

Hawkins (1973) also proposed finding PCs for the enlarged set of $(p + 1)$ variables, but he used the PCs in a rather different way from that proposed in latent root regression. The idea here is to use the PCs themselves, or rather a rotated version of them, to decide upon a suitable regression equation. Any PC with a small variance gives a relationship between y and the predictor variables whose sum of squared residuals orthogonal to the fitted plane is small. Of course, in regression, it is squared residuals in the y-direction, rather than orthogonal to the fitted plane, which are to be minimized (see Section 8.6), but the low-variance PCs can nevertheless be used to suggest low-variability relationships between y and the predictor variables. Hawkins (1973) goes further by suggesting that it may be more fruitful to look at rotated versions of the PCs, instead of the PCs themselves, in order to indicate low-variance relationships. This is done by rescaling, and then using varimax rotation (see Chapter 7) which has the effect of transforming from the PCs to a different set of orthogonal variables. These variables are, like the PCs, orthogonal linear functions of the original $(p + 1)$ variables, but their coefficients are mostly close to zero, or a long way from zero, with very few intermediate values. There is no guarantee, in general, that any of the new variables will have particularly large or particularly small variances since they are chosen by simplicity of structure of their coefficients, rather than for their variance properties. However, if only one or two of the coefficients for y are large, as should often happen with varimax rotation, then Hawkins (1973) shows that the corresponding transformed variables will have very small variances, and will therefore suggest low-variance relationships between y and the predictor variables. Other possible regression equations may be found by substitution of one subset of predictor variables in terms of another, using any low-variability relationships between predictor variables which may be suggested by the other rotated PCs.

The above technique is advocated by Hawkins (1973) and by Jeffers (1981) as a means of selecting which variables should appear in the regression equation (see the next section of this chapter), rather than as a way of directly estimating their coefficients in the regression equation, although it could be used for the latter purpose. Daling and Tamura (1970) also discussed rotation of PCs in the context of variable selection, but their PCs were for the predictor variables only.

In a later paper, Hawkins and Eplett (1982) propose another variant of latent root regression which can be used to efficiently find low-variability relationships between y and the predictor variables, and also in variable selection. This method replaces the rescaling and varimax rotation of Hawkins' earlier method by a sequence of rotations, leading to a set of relationships between y and the predictor variables which are simpler to interpret than in the previous method. This simplicity is achieved because the matrix of coefficients defining the relationships has non-zero entries only in its lower-triangular region. Despite the apparent complexity of the new method, it is also computationally simple to implement. The covariance (or correlation) matrix, $\tilde{\Sigma}$, of y and all the predictor variables, is factorized using a Cholesky factorization

$$\tilde{\Sigma} = \mathbf{DD'},$$

where $\mathbf{D}$ is lower-triangular. Then, the matrix of coefficients defining the relationships is proportional to $\mathbf{D}^{-1}$, which is also lower-triangular. To find $\mathbf{D}$ it is not necessary to calculate PCs based on $\tilde{\Sigma}$, which makes the links between the method and PCA rather more tenuous than that between PCA and latent root regression.

The next section discusses variable selection in regression using PCs, and because all three variants of latent root regression described above can be used in variable selection, they will all be discussed further in that section.

Another variation on the idea of PC regression has been used in several unpublished meteorological examples, in which a multivariate (rather than multiple) regression analysis is appropriate, i.e. where there are several dependent variables, as well as regressor variables. Here, PCA is performed on the dependent variables and, separately, on the predictor variables. A number of PC regressions are then carried out with, as usual, PCs of predictor variables in place of the predictor variables, but, in each regression, the dependent variable is now one of the high variance PCs of the original set of dependent variables. The method works reasonably well in several of the examples where it has been used (for example, Maryon (1979), which was discussed in Section 4.3), and it would be appropriate if the prediction of high-variance PCs of the dependent variables is really of interest. (In this case, another possibility is to regress PCs of the dependent variables on the original predictor variables.) However, if overall optimal prediction of linear functions of dependent variables from linear functions of predictor variables is required, then canonical correlation analysis (see Section 9.3, and Mardia et al., 1979, Chapter 10) is more suitable. Alternatively, if interpretable relationships between the original sets of dependent and predictor variables are wanted, then multivariate regression analysis (see Mardia et al., 1979, Chapter 6) may be the most appropriate technique.

Finally, the so-called PLS (partial least squares) method proposed by Wold and others (see, for example, Wold et al., 1983) provides yet another way of using PCs in regression. The method was really devised for the multivariate regression situation, but as a special case, it can be used in multiple regression. Its basic idea is to start with the PCs of the predictor variables, but then rotate them so that they become more highly correlated with the dependent variable, y. These rotated or 'tilted' PCs are then used to predict y.

8.5. Variable Selection in Regression Using Principal Components

Principal component regression, latent root regression, and other biased regression estimates keep all the predictor variables in the model, but change the estimates from least squares estimates in a way which reduces the effects of multicollinearity. As mentioned in the introductory section of this chapter, an alternative way of dealing with multicollinearity problems is to use only a subset of the predictor variables. Among the very many possible methods of selecting a subset of variables, a few use PCs.

As noted in the previous section, the procedures due to Hawkins (1973) and Hawkins and Eplett (1982) can be used in this way. Rotation of the PCs produces a large number of zero or near-zero coefficients for the rotated variables, so that low-variance relationships involving y (if such low-variance relationships exist) will include only a subset of the predictor variables which have coefficients substantially different from zero. This subset forms a plausible selection of variables which should be included in a regression model. Other low-variance relationships between the predictor variables alone may exist, again with relatively few coefficients far from zero. If such relationships exist, and involve some of the same variables as were in the relationship involving y, then substitution will lead to alternative subsets of predictor variables. Jeffers (1981) argues that in this way, it is possible to identify all good subregressions, using Hawkins' (1973) original procedure. Hawkins and Eplett (1982) demonstrate that their newer technique, using Cholesky factorization, can do even better than the earlier method. In particular, for an example which was analysed by both methods, two subsets of variables selected by the first method were shown to be inappropriate by the second.

Principal component regression and latent root regression may also be used, in an iterative manner, to select variables. Consider, first, PC regression and suppose that $\tilde{\boldsymbol{\beta}}$, given by (8.1.12), is the proposed estimator for $\boldsymbol{\beta}$. Then it is possible to test whether or not subsets of the elements of $\tilde{\boldsymbol{\beta}}$ are significantly different from zero, and those variables whose coefficients are found to be not significantly non-zero can then be deleted from the model. Mansfield et al. (1977), after a moderate amount of algebra, constructed the appropriate tests for estimators of the form (8.1.10), i.e. where the PCs deleted from the regression are restricted to be those with the smallest variances. Provided that the true coefficients of the deleted PCs are zero, and that normality assumptions are valid, then the appropriate test statistics are F-statistics, reducing to t-statistics if we consider only one variable at a time. A corresponding result will also hold for the more general form of estimator (8.1.12).

Although the variable selection procedure could stop at this stage, it may be more fruitful to use an iterative procedure, similar to that suggested by Jolliffe (1972) for variable selection in another (non-regression) context—see Section 6.3, method (i). The next step in such a procedure would be to perform a PC regression on the reduced set of variables, and then see if any further variables could be deleted from the reduced set, using the same reasoning as before. This process could be repeated, until eventually no more variables are deleted. Two variations on this iterative procedure are described by Mansfield et al. (1977). The first is a stepwise procedure which first looks for the best single variable to delete, then the best pair of variables, one of which is the best single variable, then the best triple of variables which includes the best pair, and so on. The procedure stops when the test for zero regression coefficients on the subset of excluded variables first gives a significant result. The second variation is to delete only one variable at each stage, and then recompute the PCs using the reduced set of variables, rather than allowing the deletion of several variables before the PCs are recomputed. According to Mansfield et al. (1977) this second variation gives, for several examples, an improved performance for the selected variables compared with subsets selected by the other possibilities. Only one example is described in detail in their paper, and this will be discussed further in the final section of the present chapter. In this example, Mansfield et al. (1977) adapt their method still further by discarding a few low variance PCs before attempting any selection of variables.

A similar procedure to that just described for PC regression can be constructed for latent root regression, this time leading to approximate F-statistics —see Gunst and Mason (1980, p. 339). Such a procedure has been described and illustrated by Webster et al. (1974) and Gunst et al. (1976).

Baskerville and Toogood (1982) have also suggested that the PCs which appear in latent root regression can be used to select subsets of the original predictor variables. Their procedure divides the predictor variables into four groups, on the basis of their coefficients in the PCs, where each of the groups has a different degree of potential usefulness in the regression equation. The first group of predictor variables defined by Baskerville and Toogood (1982) consists of 'isolated' variables, which are virtually uncorrelated with y and with all other predictor variables; such variables can clearly be deleted. The second and third groups contain variables which are involved in non-predictive and predictive multicollinearities respectively; those variables in the second group can usually be excluded from the regression analysis, whereas those in the third group certainly cannot. The fourth group simply consists of variables which do not fall into any of the other three groups. These variables may or may not be important in the regression, depending on the purpose of the analysis (e.g. prediction, or identification of structure) and each must be examined individually—see Baskerville and Toogood (1982) for an example.

A further possibility for variable selection is based on the idea of associating a variable with each of the first few, or the last few components, and then retaining (deleting) those variables associated with the first few (last few) PCs. This procedure was described in a different context in Section 6.3, and it is clearly essential to modify it in some way for use in a regression context. In particular, when there is not a single clear-cut choice of which variable to associate with a particular PC, the choice should be determined by looking at the strength of the relationships between the candidate variables and the dependent variable. Great care would also be necessary to avoid deletion of variables which occur in a predictive multicollinearity.

Daling and Tamura (1970) adopt a modified version of this type of approach. They first delete the last few PCs, then rotate the remaining PCs using varimax, and finally select one variable associated with each of those rotated PCs which has a 'significant' correlation with the dependent variable. The method therefore takes into account the regression context of the problem at the final stage, and the varimax rotation will increase the chances of an unambiguous choice of which variable to associate with each PC. The main drawback of the approach is in its first stage, where deletion of the low-variance PCs may discard substantial information regarding the relationship between y and the predictor variables, as was discussed in Section 8.2.

8.6. Functional and Structural Relationships

In the standard regression framework, the predictor variables are implicitly assumed to be measured without error, whereas any measurement error in the dependent variable y can be included in the error term ε. If all the variables are subject to measurement error, the problem is more complicated, even when there is only one predictor variable, and much has been written on how to estimate the so-called functional or structural relationships between the variables in such cases—see, for example, Kendall and Stuart (1979, Chapter 29).

Consider the case where there are $(p + 1)$ variables $x_0, x_1, x_2, \ldots, x_p$ which have a *linear functional relationship* (Kendall and Stuart, 1979, p. 416)

$$\sum_{j=0}^{p} \beta_j x_j = \text{const.} \tag{8.6.1}$$

between them, but which are all subject to measurement error, so that we actually have observations on $\xi_0, \xi_1, \xi_2, \ldots, \xi_p$, where

$$\xi_j = x_j + e_j, \qquad j = 0, 1, 2, \ldots, p,$$

and e_j is a measurement error term. (We have included $(p + 1)$ variables in order to keep a parallel with the case of linear regression with dependent variable y, and p predictor variables $x_1, x_2, \ldots, x_p$, but there is no reason here to treat any one variable differently from the remaining p.) On the basis of n observations on $\xi_j, j = 0, 1, 2, \ldots, p$, we wish to estimate the coefficients β_0, $\beta_1, \ldots, \beta_p$ in the relationship (8.6.1). If the e_j's are assumed to be normally distributed, and (the ratios of) their variances are known, then maximum likelihood estimation of $\beta_0, \beta_1, \ldots, \beta_p$ leads to the coefficients of the last PC from the covariance matrix of $\xi_0/s_0, \xi_1/s_1, \ldots, \xi_p/s_p$, where $s_j^2 = \text{var}(e_j)$. If there is no information about the variances of the e_j's, then no formal estimation procedure is possible, but, if it is expected that the measurement errors of all $(p + 1)$ variables are of similar variability, then a reasonable procedure is to use the last PC of $\xi_0, \xi_1, \ldots, \xi_p$.

Even if there is no formal requirement to estimate a relationship such as (8.6.1), the last few PCs are still of interest in finding near-constant linear relationships among a set of variables, as discussed in Section 3.4.

When the last PC is used to estimate a 'best-fitting' relationship between a set of $(p + 1)$ variables, we are finding the p-dimensional hyperplane for which the sum of squares of *perpendicular* distances of the observations from the hyperplane is minimized. This was, in fact, one of the objectives of Pearson's (1901) original derivation of PCs—see Property G3 in Section 3.2. By contrast, if one of the $(p + 1)$ variables, y, is a dependent variable and the remaining p are predictor variables, then the 'best-fitting' hyperplane, in the least squares sense, minimizes the sum of squares of the distances, *in the y direction*, of the observations from the hyperplane, and leads to a different relationship.

A different way of using PCs in investigating structural relationships is illustrated by Rao (1964). In his example, there are 20 variables corresponding to measurements of 'absorbance' made by a spectrophotometer, at 20 different wavelengths. There are 54 observations of the 20 variables, corresponding to nine different spectrophotometers, each used under three conditions on two separate days. The aim is to relate the absorbance measurements to wavelengths; both are subject to measurement error, so that a structural relationship, rather than straightforward regression analysis, is of interest. In this example, the first PCs, rather than the last, proved to be useful in investigating aspects of the structural relationship. Examination of the values of the 54 observations for the first two PCs identified systematic differences between spectrophotometers in the measurement errors for wavelength. Other authors have used similar, but rather more complicated ideas, based on PCs, for the same type of data. Naes (1985) refers to the problem as one of multivariate calibration, and investigates an estimate (which uses PCs) for some chemical or physical quantity, given a number of spectrophotometer measurements. Sylvestre *et al.* (1974) take as their objective the identification and estimation of mixtures of two or more overlapping curves in spectrophotometry, and again use PCs in their procedure.

8.7. Examples of Principal Components in Regression

Early examples of PC regression include those given by Kendall (1957, p. 71), Spurrell (1963) and Massy (1965). Examples of latent root regression in one form or another, and its use in variable selection, are given by Gunst *et al.* (1976), Gunst and Mason (1977b), Hawkins (1973), Baskerville and Toogood (1982) and Hawkins and Eplett (1982). In Gunst and Mason (1980, Chapter 10) PC regression, latent root regression and ridge regression are all illustrated, and can therefore be compared, for the same data set. In the present section we discuss in some detail two examples illustrating some of the techniques described in this chapter.

8.7.1. Pitprop Data

No discussion of PC regression would be complete without the example given originally by Jeffers (1967), concerning strengths of pitprops, which has since been discussed by several authors. The data consist of 14 variables which were measured for each of 180 pitprops cut from Corsican pine timber. The objective was to construct a prediction equation for one of the variables (compressive strength, y) using the values of the other 13 variables. These other 13 variables were physical measurements on the pitprops, which could be measured fairly straightforwardly without destroying the props. The variables are listed by Jeffers (1967, 1981) and the correlation matrix for all 14 variables is reproduced in Table 8.2. In his original paper, Jeffers (1967) used PC regression to predict y from the 13 variables. The coefficients for each of the PCs on the variables are given in Table 8.3. The pattern of correlations in Table 8.2 is not easy to interpret; nor is it easy to deduce the form of the first few PCs from the correlation matrix. However, Jeffers (1967) was able to interpret the first six PCs.

Also given in Table 8.3 are variances of each component, the percentage of

Table 8.2. Correlation matrix for the pitprop data.

	TOPDIAM	LENGTH	MOIST	TESTSG	OVENSG	RINGTOP	RINGBUT	BOWMAX	BOWDIST	WHORLS	CLEAR	KNOTS	DIAKNOT
LENGTH	0.954												
MOIST	0.364	0.297											
TESTSG	0.342	0.284	0.882										
OVENSG	-0.129	-0.118	-0.148	0.220									
RINGTOP	0.313	0.291	0.153	0.381	0.364								
RINGBUT	0.496	0.503	-0.029	0.174	0.296	0.813							
BOWMAX	0.424	0.419	-0.054	-0.059	0.004	0.090	0.372						
BOWDIST	0.592	0.648	0.125	0.137	-0.039	0.211	0.465	0.482					
WHORLS	0.545	0.569	-0.081	-0.014	0.037	0.274	0.679	0.557	0.526				
CLEAR	0.084	0.076	0.162	0.097	0.091	-0.036	-0.113	0.061	0.085	-0.319			
KNOTS	-0.019	-0.036	0.220	0.169	-0.145	0.024	-0.232	-0.357	-0.127	-0.368	0.029		
DIAKNOT	0.134	0.144	0.126	0.015	-0.208	-0.329	-0.424	-0.202	-0.076	-0.291	0.007	0.184	
STRENGTH	-0.419	-0.338	-0.728	-0.543	0.247	0.117	0.110	-0.253	-0.235	-0.101	-0.055	-0.117	-0.153

Table 8.3. Principal component regression for the pitprop data: coefficients, variances, regression coefficients and t-statistics for each component.

						Principal component							
	1	2	3	4	5	6	7	8	9	10	11	12	13
x_1	-0.40	0.22	-0.21	-0.09	-0.08	0.12	-0.11	0.14	0.33	-0.31	0.00	0.39	-0.57
x_2	-0.41	0.19	-0.24	-0.10	-0.11	0.16	-0.08	0.02	0.32	-0.27	-0.05	-0.41	0.58
x_3	-0.12	0.54	0.14	0.08	0.35	-0.28	-0.02	0.00	-0.08	0.06	0.12	0.53	0.41
x_4	-0.17	0.46	0.35	0.05	0.36	-0.05	0.08	-0.02	-0.01	0.10	-0.02	-0.59	-0.38
x_5	-0.06	-0.17	0.48	0.05	0.18	0.63	0.42	-0.01	0.28	-0.00	0.01	0.20	0.12
x_6	-0.28	-0.01	0.48	-0.06	-0.32	0.05	-0.30	0.15	-0.41	-0.10	-0.54	0.08	0.06
x_7 Coefficients	-0.40	-0.19	0.25	-0.07	-0.22	0.00	-0.23	0.01	-0.13	0.19	0.76	-0.04	0.00
x_8	-0.29	-0.19	-0.24	0.29	0.19	-0.06	0.40	0.01	-0.35	-0.08	0.03	-0.05	0.02
x_9	-0.36	0.02	-0.21	0.10	-0.10	0.03	0.40	0.64	-0.38	-0.06	-0.05	0.05	-0.06
x_{10}	-0.38	-0.25	-0.12	-0.21	0.16	-0.17	0.00	-0.70	0.27	0.71	-0.32	0.06	0.00
x_{11}	0.01	0.21	-0.07	0.80	-0.34	0.18	-0.14	-0.01	0.15	0.34	-0.05	0.00	-0.01
x_{12}	0.12	0.34	0.09	-0.30	-0.60	-0.17	0.54	0.21	0.08	0.19	0.05	0.00	0.00
x_{13}	0.11	0.31	-0.33	-0.30	0.08	0.63	-0.16	0.11	-0.38	0.33	0.04	0.01	-0.01
Variance	4.22	2.38	1.88	1.11	0.91	0.82	0.58	0.44	0.35	0.19	0.05	0.04	0.04
% of total variance	32.5	18.3	14.4	8.5	7.0	6.3	4.4	3.4	2.7	1.5	0.4	0.3	0.3
Regression coefficient γ_k	0.13	-0.37	0.13	-0.05	-0.39	0.27	-0.24	-0.17	0.03	0.00	-0.12	-1.05	0.00
t-value	6.86	14.39	4.38	1.26	9.23	6.19	4.50	2.81	0.46	0.00	0.64	5.26	0.01

total variation accounted for by each component, the coefficients γ_k in a regression of y on the PCs, and the values of t-statistics measuring the importance of each PC in the regression.

Judged solely on the basis of size of variance it would appear that the last three, or possibly four, PCs could be deleted from the regression. However, looking at values of γ_k and the corresponding t-statistics, it can be seen that the twelfth component is relatively important as a predictor of y, despite the fact that it accounts for only 0.3% of the total variation in the regressor variables. Jeffers (1967) only retained the first, second, third, fifth and sixth PCs in his regression equation, whereas Mardia et al. (1979, p. 246) suggest that the seventh, eighth and twelfth PCs should also be included.

This example has also been used by various authors to illustrate techniques of variable selection and some of the results are given in Table 8.4. Jeffers (1981) has used Hawkins' (1973) variant of latent root regression to select subsets of five, six or seven regressor variables. After varimax rotation, only one of the rotated components has a substantial coefficient for compressive strength, y. This rotated component has five other variables which have large coefficients, and it is suggested that these should be included in the regression equation for y; two further variables with moderate coefficients might also be included. One of the five variables definitely selected by this method is quite difficult to measure, and one of the other rotated components suggests that it can be replaced by another, more readily measured, variable. However, this substitution causes a substantial drop in the squared multiple correlation for the five-variable regression equation, from 0.695 to 0.581.

Mansfield et al. (1977) used an iterative method based on PC regression, and described above in Section 8.5, to select a subset of variables for these data. The procedure is fairly lengthy since only one variable is deleted at each iteration, but the F-criterion used to decide whether to delete an extra variable jumps from 1.1 to 7.4 between the fifth and sixth iterations, giving a clear-cut decision to delete five variables, i.e. retain eight variables. As can be seen from Table 8.4, this eight-variable subset has a large degree of overlap with the subsets found by Jeffers (1981).

Jolliffe (1973) also found subsets of the 13 variables, using various methods, but the variables in this case were chosen to reproduce the relationships between the regressor variables, rather than to predict y as well as possible. McCabe (1982), using a technique related to PCA (see Section 6.3), and with a similar purpose to Jolliffe's (1973) methods, chose subsets of various sizes. McCabe's subsets are the best few with respect to a single criterion, whereas Jolliffe gives the single best subset, but for several different methods. The best subsets due to Jolliffe (1973) and McCabe (1982) have considerable overlap with each other, but there are substantial differences from the subsets of Jeffers (1981) and Mansfield et al. (1977). This reflects the different aims of the different selection methods. It shows again that substantial variation within the set of regressor variables does not necessarily imply any relationship with y and, conversely, that variables having little correlation with the first few PCs can still be important in predicting y.

Table 8.4. Variable selection using various techniques on the pitprop data. (Each row corresponds to a selected subset with × denoting a selected variable.)

	Variables												
	1	2	3	4	5	6	7	8	9	10	11	12	13
Five variables													
Jeffers (1981)	×		×			×	×	×					
	×					×	×	×					×
McCabe (1982)				×					×		×	×	×
				×	×				×		×	×	
Six variables													
Jeffers (1981)	×		×			×	×	×	×				
McCabe (1982)				×	×			×	×		×		×
Jolliffe (1973)			×		×					×	×	×	×
McCabe (1982)		×	×		×				×		×	×	
Jolliffe (1973)			×		×			×			×	×	×
Eight variables													
Mansfield et al. (1977)	×		×	×	×	×	×	×			×		

Table 8.5. Variables used in household formation example.

No.	Description	No.	Description
1.	Population in non-private establishments	17.	Ratio households to rateable units
2.	Population age 0–14	18.	Domestic rateable value (£) per head
3.	Population age 15–44	19.	Rateable units with rateable value < £100
4.	Population age 60/65 +	20.	Students age 15 +
5.	Females currently married	21.	Economically active married females
6.	Married males 15–29	22.	Unemployed males seeking work
7.	Persons born ex UK	23.	Persons employed in agriculture
8.	Average population increase per annum (not births and deaths)	24.	Persons employed in mining and manufacturing
9.	Persons moved in previous 12 months	25.	Males economically active or retired in socio-economic group 1, 2, 3, 4, 13
10.	Households in owner occupation		
11.	Households renting from Local Authority	26.	Males economically active or retired in socio-economic group 5, 6, 8, 9, 12, 14
12.	Households renting private unfurnished	27.	With degrees (excluding students with degree)
13.	Vacant dwellings		
14.	Shared dwellings	28.	Economically active males socio-economic group 3, 4
15.	Households over one person per room		
16.	Households with all exclusive amenities	29.	Average annual total income (£) per adult

8.7.2. Household Formation Data

This example uses part of a data set which arose in a study of household formation. The subset of data used here has 29 demographic variables measured, in 1971, for 168 local government areas in England and Wales. The variables are listed in Table 8.5. All variables, except numbers 17, 18 and 29, are expressed as numbers per 1000 of population; precise definitions of each variable are given in Appendix B of Bassett et al. (1980).

Although this was not the purpose of the original project, the objective considered here will be to predict the final variable (average annual total income per adult) from the other 28. This objective is a useful one, since information on income is often difficult to obtain accurately, and predictions from other, more readily available, variables would be valuable. The results presented below were given by Garnham (1979) in an unpublished M.Sc. dissertation, and further details of the regression analysis are given in that source. A full description of the project from which the data are taken is available in Bassett et al. (1980). Most regression problems with as many as 28 regressor variables will have multicollinearities, and the current example is no exception. Looking at the list of variables in Table 8.5 it is clear, even without detailed definitions, that there are groups of variables which are likely to be highly correlated. For example, several variables relate to type of household, whereas another group of variables considers rates of employment in various types of job. Table 8.6, giving the eigenvalues of the correlation matrix, confirms that there are multicollinearities; some of the eigenvalues are very small.

Table 8.6. Eigenvalues of the correlation matrix for the household formation data.

PC number	1	2	3	4	5	6	7	8	9	10	11	12	13	14
Eigenvalue	8.62	6.09	3.40	2.30	1.19	1.06	0.78	0.69	0.58	0.57	0.46	0.36	0.27	0.25
Order of importance in predicting y	1	4	2	8	9	3	13	22	20	5	11	15	24	23
PC number	15	16	17	18	19	20	21	22	23	24	25	26	27	28
Eigenvalue	0.24	0.21	0.18	0.14	0.14	0.10	0.10	0.07	0.07	0.05	0.04	0.03	0.02	0.003
Order of importance in predicting y	17	25	16	10	7	21	28	6	18	12	14	27	19	26

Consider now PC regression, and some of the strategies which could be used to select a subset of PCs to be included in the regression. Deleting components with small variance, with a cutoff of about $l^* = 0.10$ would mean that between seven and nine components could be left out. Sequential deletion of PCs with the smallest variances, using t-statistics at each stage suggests that only six PCs can be deleted. However, from the point of view of R^2, the squared multiple correlation coefficient, deletion of eight or more might be acceptable; R^2 is 0.874 for the full model including all 28 variables, and it is reduced to 0.865, 0.851 respectively when five and eight components are deleted.

It is interesting to examine the ordering of size of correlations (or equivalently, the individual t-values) between y and the PCs, which is also given in Table 8.6. It is seen that those PCs with small variance do not necessarily have small correlations with y. The 18th, 19th and 22nd in size of variance are in the first ten in order of importance for predicting y; in particular, the 22nd PC with variance 0.07, has a highly significant t-value, and should almost certainly be retained.

An approach using stepwise deletion based solely on the size of correlation between y and each PC produces, because of the zero correlations between PCs, the subset whose value of R^2 is maximized, for any given subset size. Far fewer PCs need to be retained using this approach than the 20 to 23 indicated when only small-variance components are rejected. In particular, if the 10 PCs are retained which best predict y, then R^2 is 0.848, compared with 0.874 for the full model and 0.851 using the first 20 PCs. It would appear that, again, a strategy based solely on size of variance is unsatisfactory.

The two 'weak MSE' criteria described in Section 8.2 were also tested in a limited way, on these data. Because of computational constraints it was not possible to find the overall 'best' subset, M, so a stepwise approach was adopted, deleting PCs according to either size of variance, or correlation with y. The first criterion selected 22 PCs when selection was based on size of variance, but only 6 PCs when correlation with y was the basis for stepwise selection. The corresponding results for the second (predictive) criterion were 24 and 12 PCs respectively. It is clear, once again, that selection based solely on order of size of variance is unwise.

The alternative approach of Lott (1973) was also investigated for these data, in a stepwise manner, using correlation with y to determine order of selection, with the result that $\bar{R}^2$ was maximized for 19 PCs. This is a substantially larger number than was indicated by those other methods which used correlation with y to define order of selection, and, given the concensus from the other methods, may suggest that Lott's (1973) method is not ideal.

When PCs are found for the augmented set of variables, including y and all the regressor variables, as required for latent root regression, there is remarkably little change in the PCs, apart from the addition of an extra one. All of the coefficients on the regressor variables are virtually unchanged, and the PCs which have largest correlation with y are in very nearly the same order as in the PC regression.

It may be of more interest to select a subset of variables, rather than a subset of PCs, to be included in the regression, and this has also been attempted, using various methods, for the household formation data. Variable selection based on PC regression, deleting just one variable at a time before recomputing the PCs as suggested by Mansfield *et al.* (1977), indicated that only 12, and possibly fewer, variables need to be retained. R^2 for the 12 variable subset given by this method is 0.862, and it has only dropped to 0.847 for the eight-variable subset (cf. 0.876 for the full model and 0.851 using the first 20 PCs in the regression). Other variable selection methods, described by Jolliffe (1972), and in Section 6.3, were also tried, but these did not produce quite such good results as the Mansfield *et al.* method. This was not surprising since, as noted in the previous example, they are not specifically tailored for variable selection in the context of regression. However, they did confirm that only eight to ten variables are really necessary in order to provide an adequate prediction of 'income' for these data.

CHAPTER 9

Principal Components Used with Other Multivariate Techniques

Principal component analysis is often used as a dimension-reducing technique within some other type of analysis. For example, Chapter 8 described the use of PCs as regressor variables in a multiple regression analysis. The present chapter discusses three multivariate techniques, namely discriminant analysis, cluster analysis and canonical correlation analysis; for each of these three techniques, examples are given in the literature which use PCA as a dimension-reducing technique.

Discriminant analysis is concerned with data in which each observation comes from one of several well-defined groups or populations. Assumptions are made about the structure of the populations, and the main objective is to construct rules for assigning future observations to one of the populations, so as to minimize the probability of misclassification, or some similar criterion. The use of PCA as a preliminary to discriminant analysis is discussed in Section 9.1. In addition, there is a brief description of a new discriminant technique which uses PCs, and discussion of a link between PCA and a standard form of discriminant analysis.

Cluster analysis is perhaps the most popular context in which PCs are derived in order to reduce dimensionality prior to the use of a different multivariate technique. Like discriminant analysis, cluster analysis deals with data sets in which the observations are to be divided into groups. However, in cluster analysis, little or nothing is known a priori about the groups, and the objective is to divide the given observations into groups or clusters in a 'sensible' way. Within cluster analysis, PCs can be employed in two separate rôles, either to construct distance measures or to provide a graphical representation of the data; the latter is often called ordination or scaling—see also Section 5.1—and is useful in detecting or verifying a cluster structure. Both rôles are described, and illustrated with examples, in Section 9.2. Also discussed in Section 9.2 is the idea of clustering variables, rather than observations, and a connection between PCA and this idea is described.

The third, and final, multivariate technique discussed in this chapter, in Section 9.3, is canonical correlation analysis. This technique is appropriate when the vector of random variables, $\mathbf{x}$, is divided into two parts, $\mathbf{x}_{p_1}, \mathbf{x}_{p_2}$, and the objective is to find pairs of linear functions, of $\mathbf{x}_{p_1}$ and $\mathbf{x}_{p_2}$ respectively, such that the correlation between the linear functions within each pair is maximized. In this case, the replacement of $\mathbf{x}_{p_1}, \mathbf{x}_{p_2}$ by some or all of the PCs of $\mathbf{x}_{p_1}, \mathbf{x}_{p_2}$ respectively, has been suggested in the literature.

9.1. Discriminant Analysis

In discriminant analysis, observations may be taken from any of $g \geq 2$ populations or groups. Assumptions are made regarding the structure of these groups, i.e. the random vector, $\mathbf{x}$, associated with each observation is assumed to have a particular (partly or fully specified) distribution, depending on its group membership, and information may also be available about the overall relative frequencies of occurrence of each group. In addition, there is usually available a set of data $\mathbf{x}_1, \mathbf{x}_2, \ldots, \mathbf{x}_n$ (the training set) for which the group membership of each observation is known. Based on the assumptions about group structure (and on the training set if one is available), rules are constructed for assigning future observations to one of the g groups in some 'optimal' way, e.g. so as to minimize the probability or cost of misclassification.

The best-known form of discriminant analysis is when there are only two populations, and $\mathbf{x}$ is assumed to have a multivariate normal distribution which differs with respect to the mean, but not the covariance matrix, between the two populations. If the means, $\boldsymbol{\mu}_1, \boldsymbol{\mu}_2$, and the common covariance matrix, $\boldsymbol{\Sigma}$, are known, then the optimal rule (according to several different criteria) is based on the linear discriminant function $\mathbf{x}'\boldsymbol{\Sigma}^{-1}(\boldsymbol{\mu}_1 - \boldsymbol{\mu}_2)$. If $\boldsymbol{\mu}_1, \boldsymbol{\mu}_2, \boldsymbol{\Sigma}$ are estimated from a 'training set' by $\bar{\mathbf{x}}_1, \bar{\mathbf{x}}_2, \mathbf{S}$ respectively, then a rule based on the sample linear discriminant function $\mathbf{x}'\mathbf{S}^{-1}(\bar{\mathbf{x}}_1 - \bar{\mathbf{x}}_2)$ is often used. There are many other varieties of discriminant analysis (Lachenbruch, 1975), depending on the assumptions made regarding the population structure, and much recent work has been done, in particular, on discriminant analysis for discrete data, and on non-parametric approaches (Goldstein and Dillon, 1978; Hand, 1982).

The most obvious way of using PCA in a discriminant analysis is to replace $\mathbf{x}$ by its PCs. Clearly, no advantage will be gained if all p of the PCs are used, so typically only the first m (high variance) PCs will be retained. These m PCs can then be substituted instead of $\mathbf{x}$ in the derivation of a discriminant rule, thus reducing the dimensionality of the problem. In addition, the first two PCs, if they account for a high proportion of the variance, can be used to provide a two-dimensional graphical representation of the data, showing how good, or otherwise, is the separation between the groups.

The first point to be clarified is exactly what is meant by the PCs of **x** in the context of discriminant analysis. A common assumption in many forms of discriminant analysis is that the covariance matrix is the same for all groups, and the PCA is therefore done on an estimate of this common *within-group* covariance (or correlation) matrix.

Unfortunately, this procedure may be unsatisfactory for two reasons. First, the within-group covariance matrix may be different for different groups. Methods for comparing PCs from different groups are discussed in Section 11.5, and later in the present section we describe a technique which uses PCs to discriminate between populations when equal covariance matrices are not assumed. For the moment, however, we make the equal-covariance assumption.

The second, more serious, problem encountered in using PCs, based on a common within-group covariance matrix, to discriminate between groups, is that there is no guarantee that the separation between groups will be in the direction of the high-variance PCs. This point is illustrated diagramatically in Figures 9.1 and 9.2 for two variables. In both figures the two groups are well separated but, in the first, the separation is in the direction of the first PC (i.e. parallel to the major axis of within-group variation), whereas, in the second, the separation is orthogonal to this direction. Thus, the first few PCs will only be useful for discriminating between groups if within- and between-group variation have the same dominant directions. If this does not occur (and in general there is no particular reason for it to do so) then omitting the low-variance PCs may actually throw away most of the information in **x** concerning between-group variation.

The problem is essentially the same one which arises in PC regression where, as discussed in Section 8.2, it is inadvisable to look only at high-variance PCs, since the low-variance PCs can also be highly correlated with the dependent variable. That the same problem arises in both multiple regression and discriminant analysis is hardly surprising, since linear discriminant analysis can be viewed as a special case of multiple regression, in which the dependent variable is a dummy variable defining group membership.

Chang (1983) gives results which are relevant to the use of PCs for a special case of discriminant analysis. It is well known that, for two completely specified normal populations, differing only in mean, the probability of misclassification using the linear discriminant function is a monotonically decreasing function of the squared Mahalanobis distance, δ^2, between the two populations, which is defined as

$$\delta^2 = (\mathbf{\mu}_1 - \mathbf{\mu}_2)'\mathbf{\Sigma}^{-1}(\mathbf{\mu}_1 - \mathbf{\mu}_2). \qquad (9.1.1)$$

(In equation (5.3.5) of Section 5.3, we defined Mahalanobis distance *between two observations in a sample*. There is an obvious similarity between (5.3.5) and the definition given in (9.1.1) for Mahalanobis distance *between two*

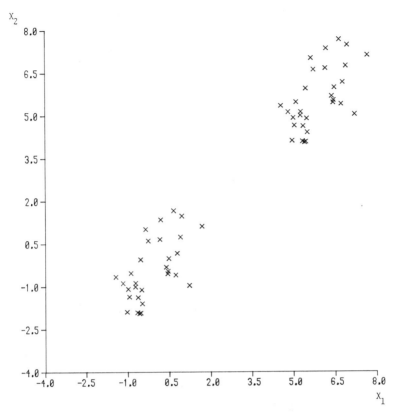

Figure 9.1. Two data sets whose direction of separation is the same as that of the first (within-group) PC.

populations. Further modifications give Mahalanobis distance between two samples (see equation (9.1.2) below), between an observation and a sample mean (see Section 10.1, below equation (10.1.2)), or between an observation and a population mean.) Thus, if we take a subset of the original p variables, then the discriminatory power of the subset can be measured by the Mahalanobis distance between the two populations in the subspace defined by the subset of variables. Chang (1983) shows that the Mahalanobis distance based on the kth PC is a monotonic increasing function of $[\alpha'_k(\mu_1 - \mu_2)]^2/\lambda_k$, where α_k, λ_k are, as usual, the vector of coefficients in the kth PC and the variance of the kth PC respectively. Therefore, the PC with the largest discriminatory power is the one which maximizes $[\alpha'_k(\mu_1 - \mu_2)]^2/\lambda_k$, and this will not necessarily correspond to the first PC, which maximizes λ_k. Indeed, if α'_1 is orthogonal to $(\mu_1 - \mu_2)$, as in Figure 9.2, then the first PC has no discriminatory power at all. Chang (1983) shows with a real example that low-variance PCs can be important discriminators in practice, and he also demonstrates that a

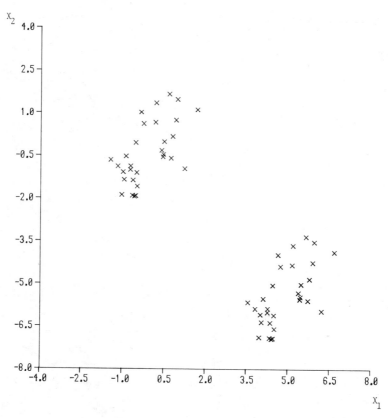

Figure 9.2. Two data sets whose direction of separation is orthogonal to that of the first (within-group) PC.

change of scaling for the variables (e.g. in going from a covariance to a correlation matrix) can change the relative importance of the PCs.

The fact that separation between populations may be in the directions of the last few PCs does not mean that PCs should not be used at all in discriminant analysis. They can still provide a reduction of dimensionality, and, as in regression, their orthogonality implies that, in linear discriminant analysis, each PC's contribution can be assessed independently. This is an advantage compared to using the original variables **x**, where the contribution of one of the variables depends on which other variables are also used in the analysis (unless all elements of **x** are uncorrelated). The main point to bear in mind in using PCs in discriminant analysis is that the best subset of PCs does not necessarily simply consist of those with the largest variances. It is easy to see, because of orthogonality, which of the PCs are best at discriminating between the populations. However, as in regression, some caution is advisable in using PCs with very low variances, because at least some of the estimated coefficients in the discriminant function will have large variances if low variance PCs are included. Many of the comments made in Section 8.2, regarding strategies for selecting PCs in regression, are also relevant in linear discriminant analysis.

So far in this section it has been assumed that PCs in discriminant analysis are based on the pooled within-group covariance matrix. This approach is valid for many types of discriminant analysis where the covariance structure is assumed to be the same for all populations. However, it is not always realistic to make this assumption, in which case some form of non-linear discriminant analysis may be necessary. Alternatively, the convenience of looking only at linear functions of $\mathbf{x}$ can be kept by computing PCs separately for each population. In a number of papers (see, for example, Wold, 1976; Wold et al., 1983), Wold and others have described a method for discriminating between populations which adopts this approach. The method, called SIMCA (Soft Independent Modelling of Class Analogy!), does a separate PCA for each group, and retains sufficient PCs in each to account for most of the variation within that group. The number of PCs retained will typically be different for different populations. To classify a new observation, the distance of the observation from the hyperplane defined by the retained PCs is calculated for each population. The square of this distance for a particular population is simply the sum of squares of the values of the omitted PCs for that population, evaluated for the observation in question. The same type of quantity is also used for detecting outliers (see Section 10.1, equation (10.1.1)).

If future observations are to be assigned to one and only one population, then assignment will be to the population for which the distance is minimized. Alternatively, a firm decision may not be required and, if all the distances are large enough, the observation may be left unassigned. Since it is not close to any of the existing groups, it may be an outlier, or come from a new group about which there is currently no information. Conversely, if the groups are not all well separated, some future observations may have small distances from more than one population. In such cases, it may again be undesirable to decide on a single possible class; instead two or more groups may be listed as possible 'homes' for the observation.

According to Wold et al. (1983), SIMCA works with as few as five objects from each population and there is no restriction on the number of variables. This is apparently important in many chemical problems where the number of variables can greatly exceed the number of observations. SIMCA can also cope with situations where one class is very diffuse, simply consisting of all observations which do not belong in one of a number of well-defined classes. Most standard discrimination techniques would break down in such situations.

SIMCA calculates PCs separately within each group, compared with the more usual practice of finding PCs for a pooled within-group covariance matrix. A third possibility, which is mentioned by Rao (1964), is to ignore the group structure and calculate an overall covariance matrix based on the raw data. This seems to have been done by Mager (1980b) in the course of using PCA in a non-linear form of discriminant analysis. If the between-group variation is much larger than within-group variation, then the first few PCs for the overall covariance matrix will define directions in which there are large between-group differences. Such PCs would therefore seem more useful than those based on within-group covariance matrices. The approach is not to be recommended, however, since it will work only if between-group variation dominates within-group variation. Furthermore, if orthogonal linear functions are required which discriminate as well as possible between the groups, then so-called canonical variables, rather than these PCs, are the appropriate functions to use. Canonical variables are defined as $\gamma_1' \mathbf{x}, \gamma_2' \mathbf{x}, \ldots,$ $\gamma_{g-1}' \mathbf{x}$ where $\gamma_k' \mathbf{x}$ maximizes the ratio of between- to within-group variance of $\gamma' \mathbf{x}$, subject to being uncorrelated with $\gamma_1' \mathbf{x}, \gamma_2' \mathbf{x}, \ldots, \gamma_{k-1}' \mathbf{x}$. For more details see, for example, Lachenbruch (1975, p. 66) and Mardia et al. (1979, Section 12.5).

To conclude this section, we note a relationship between PCA and canonical discriminant analysis, via principal co-ordinate analysis (see Section 5.2), which was described by Gower (1966). Suppose that a principal co-ordinate analysis is done on a distance matrix whose elements are Mahalanobis distances between the samples from the g populations. These distances are defined as the square roots of

$$\delta_{hi}^2 = (\bar{\mathbf{x}}_h - \bar{\mathbf{x}}_i)' \mathbf{S}^{-1} (\bar{\mathbf{x}}_h - \bar{\mathbf{x}}_i); \qquad h, i = 1, 2, \ldots, g. \qquad (9.1.2)$$

Gower (1966) then showed that the configuration found in $m \ (< g)$ dimensions is the same as that provided by the first m canonical variables; furthermore, the same results may be found from a PCA with $\mathbf{X}'\mathbf{X}$ replaced by $(\bar{\mathbf{X}}\mathbf{W})'(\bar{\mathbf{X}}\mathbf{W})$, where $\mathbf{W}\mathbf{W}' = \mathbf{S}^{-1}$ and $\bar{\mathbf{X}}$ is the $(g \times p)$ matrix whose hth row gives the sample means of the p variables for the hth population, $h = 1,$ $2, \ldots, g$. Because of this connection between PCs and canonical variables, Mardia et al. (1979, p. 344) refer to canonical discriminant analysis as the analogue for grouped data of PCA for ungrouped data.

9.2. Cluster Analysis

In cluster analysis, it is required to divide a set of observations into groups or clusters in such a way that most pairs of observations which are placed in the same group are more similar to each other than are pairs of observations which are placed into two different clusters. In some circumstances, it may be expected or hoped that there is a clear-cut group structure underlying the data, so that each observation comes from one of several distinct populations, as in discriminant analysis. The objective then is to determine this group structure, where, in contrast to discriminant analysis, there is little or no prior information about the form which the structure takes. Cluster analysis can also be useful when there is no clear group structure in the data. In this case, it may still be desirable to subdivide, or *dissect* (using the terminology of Kendall (1966)) the observations into relatively homogeneous groups since observations within the same group may be sufficiently similar to be treated identically for the purpose of some further analysis, whereas this would be impossible for the whole, heterogeneous, data set. There is a very large number of possible methods of cluster analysis, and several books have appeared on the subject, e.g. Aldenderfer and Blashfield (1984), Everitt (1980), Gordon (1981) and Hartigan (1975); most methods can be used either for detection of clear-cut groups or for dissection.

The majority of cluster analysis techniques require a measure of similarity or dissimilarity between each pair of observations, and PCs have been used quite extensively in the computation of one type of dissimilarity. If the p variables which are measured for each observation are quantitative and in similar units, then an obvious measure of dissimilarity between two observations is the Euclidean distance between the observations, in the p-dimensional space defined by the variables. If the variables are measured in non-compatible units, then each variable can be standardized by dividing by its standard deviation, and an arbitrary, but obvious, measure of dissimilarity is then the Euclidean distance between a pair of observations in the p-dimensional space defined by the standardized variables.

Suppose that a PCA is done, based on the covariance or correlation matrix, and that m $(<p)$ PCs account for most of the variation in $\mathbf{x}$. A possible alternative dissimilarity measure is the Euclidean distance between a pair of observations in the m-dimensional subspace defined by the first m PCs; such dissimilarity measures have been used in several published studies, e.g. Jolliffe *et al.* (1980). There is often no real advantage in using this measure, rather than the Euclidean distance in the original p-dimensional space, since the Euclidean distance calculated using all p PCs from the covariance matrix is identical to that calculated from the original variables. Similarly, the distance calculated from all p PCs for the correlation matrix is the same as that calculated from the p standardized variables. Using m, instead of p, PCs simply provides an approximation to the original Euclidean distance, and the extra calculation involved in finding the PCs far outweighs any saving which results from using m, instead of p, variables in computing the distance. However, if, as in Jolliffe *et al.* (1980), the PCs are being calculated in any case, the reduction from p to m variables may be worthwhile.

In calculating Euclidean distances, the PCs have the usual normalization, so that var$(\mathbf{a}_k'\mathbf{x}) = l_k, k = 1, 2, \ldots, p$ and $l_1 \geq l_2 \geq \cdots \geq l_p$, using the notation of Section 3.1. As an alternative, a distance can be calculated based on PCs which have been renormalized so that each PC has the same variance. This renormalization is discussed further in the context of outlier detection in the next chapter. In the present setting, where the objective is the calculation of a dissimilarity measure, its use is based on the following idea. Suppose that one of the original variables is almost independent of all the others, but that several of the remaining variables are measuring essentially the same property as each other. Euclidean distance will then give more weight to this property than to the property described by the 'independent' variable. If it is thought desirable to give equal weight to each property then this can be achieved by finding the PCs and then giving equal weight to each of the first m PCs.

To see that this works consider a simple example in which four meteorological variables were measured. Three of the variables are temperatures, namely air temperature, sea surface temperature and dewpoint, and the fourth is the height of the cloudbase. The first three variables are highly correlated with each other, but nearly independent of the fourth. For a sample of 30 measurements of these variables, a PCA based on the correlation matrix gave a first PC, with variance 2.95, which was a nearly-equally-weighted average of the three temperature variables. The second PC, with variance 0.99 was dominated by cloudbase height, and together the first two PCs accounted for 98.5% of the total variation in the four variables.

Euclidean distance based on the first two PCs gives a very close approximation to Euclidean distance based on all four variables, but it gives roughly three times as much weight to the first PC as to the second. Alternatively, if the first two PCs are renormalized to have equal weight, this implies that we are treating the *one* measurement of cloudbase height as being equally important as the *three* measurements of temperature.

In general, if Euclidean distance is calculated using all p renormalized PCs, then this is equivalent to calculating the Mahalanobis distance for the original variables (see Section 10.1, below equation (10.1.2), for a proof of the corresponding property for Mahalanobis distances of observations from sample means, rather than between pairs of observations). Mahalanobis distance is yet another plausible dissimilarity measure, which takes into account the variances and covariances between the elements of **x**.

Regardless of the similarity or dissimilarity measure adopted, PCA has a further use in cluster analysis, namely to provide a two-dimensional representation of the observations—see also Section 5.1. Such a two-dimensional representation can give a simple visual means of either detecting or verifying the existence of clusters (as noted by Rao (1964)), provided that most of the variation, and in particular the between-cluster variation, falls in the two-dimensional subspace defined by the first two PCs.

Of course the same problem can arise as in discriminant analysis, namely that the between-cluster variation may be in directions other than those of the first two PCs, even if these two PCs account for nearly all of the total variation. However, this behaviour is generally less likely in cluster analysis since the PCs are calculated for the whole data set, not within-groups; as pointed out in the previous section, if between-cluster variation is much greater than within-cluster variation, such PCs will often successfully reflect the cluster structure. It is, in any case, typically impossible to calculate within-group PCs in cluster analysis since the group structure is usually completely unknown *a priori*. Chang's (1983) criticism of the use of the first few PCs in '*clustering*' is somewhat overstated since his PCs are based on within-group covariances and correlations, so that his results are more relevant to *discriminant analysis*, as discussed in the previous section.

Tukey and Tukey (1981) argue that there are often better directions than PCs in which to view the data in order to 'see' structure such as clusters, and 'projection-pursuit' methods have been devised to find appropriate directions—see also Sibson (1984). However, the examples discussed below will illustrate that plots with respect to the first two PCs can give suitable two-dimensional representations on which to view the cluster structure, if a clear structure exists. Furthermore, in the case where there is no clear structure, but it is required to dissect the data using cluster analysis, there can be no real objection to the use of a plot with respect to the first two PCs. If we wish to view the data in two dimensions in order to see whether a set of clusters given by some procedure 'looks sensible', then the first two PCs give the best possible representation in two dimensions in the sense defined by Property G3 of Section 3.2.

Before looking at examples of the uses, just described, of PCA in cluster analysis, we describe briefly a rather different way in which cluster analysis can be used, and a 'connection' with PCA. So far, we have discussed cluster analysis on observations, or individuals, but in some circumstances it is desirable to divide variables, rather than observations, into groups. In fact, by far the earliest book on cluster analysis (Tryon, 1939) was concerned with this type of application. Provided that a suitable measure of similarity between variables can be defined—the correlation coefficient is an obvious candidate—methods of cluster analysis used for observations can be readily adapted for variables.

The connection with PCA is that when the variables fall into well-defined clusters, then, as discussed in Section 3.8, there will then be one high-variance PC and, except in the case of 'single-variable' clusters, one or more low-variance PCs, associated with each cluster of variables. Thus, PCA will identify the presence of clusters among the variables, and can be thought of as a competitor to standard cluster analysis of variables.

Identifying clusters of variables may be of general interest in investigating the structure of a data set, but, more specifically, if we wish to reduce the number of variables without sacrificing too much information, then we could retain one variable from each cluster. This is essentially the idea behind the variable selection techniques, based on PCA, which were described in Section 6.3.

9.2.1. Examples

Only one example will be described in detail here, although a number of other examples which have appeared elsewhere will be discussed briefly. In many of the published examples where PCs have been used in conjunction

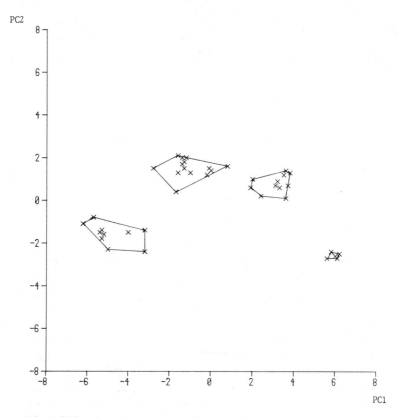

Figure 9.3. Aphids: plot with respect to first two PCs showing four groups corresponding to species.

with cluster analysis, there is no clear-cut cluster structure, and cluster analysis has been used as a dissection technique. An exception is the well-known example given by Jeffers (1967), which was discussed in the context of variable selection in Section 6.4.1. The data consist of 19 variables measured on 40 aphids, and, when the 40 observations are plotted with respect to the first two PCs, there is a strong suggestion of four distinct groups—see Figure 9.3, on which convex hulls (see Section 5.1) have been drawn around the four suspected groups. It is likely that the four groups indicated on Figure 9.3 correspond to four different species of aphids; these four species cannot be readily distinguished using only one variable at a time, but the plot with respect to the first two PCs clearly distinguishes the four populations.

The first example of Section 5.1.1 dealing with seven physical measurements on 28 students, also shows (in Figure 5.1) how a plot with respect to the first two PCs can distinguish two groups, in this case men and women. There is, unlike the aphid data, a small amount of overlap between groups and if the PC plot is used to identify, rather than verify, a cluster structure, then it is likely that some misclassification between sexes will occur.

In the situation where cluster analysis is used for dissection, the aim of a two-dimensional plot with respect to the first two PCs will almost always be to verify that a given dissection 'looks' reasonable, rather than to attempt to identify clusters. An early example of this type of use was given by Moser and Scott (1961), in their Figure 2. The PCA in their study, which has already been mentioned in Section 4.2, was a stepping stone on the way to a cluster analysis of 157 British towns based on 57 variables. The PCs were used both in the construction of a distance measure, and as a means of displaying the clusters in two dimensions.

Principal components were used in cluster analysis in a similar manner in other examples discussed in Section 4.2, details of which can be found in Jolliffe *et al.* (1980, 1982a, 1986), Imber (1977) and Webber and Craig (1978). Each of these studies was concerned with demographic data, as is the example which is now described in detail.

Demographic Characteristics of English Counties

In an unpublished undergraduate dissertation, Stone (1984), considered a cluster analysis of 46 English counties. For each county there were 12 demographic variables, which are listed in Table 9.1. The objective of Stone's analysis, namely dissection of local authority areas into clusters, was basically the same as that in other analyses by Imber (1977), Webber and Craig (1978) and Jolliffe *et al.* (1986), but these various analyses differ in the variables used and in the local authorities considered. For example, Stone's list of variables is shorter than those of the other analyses, although it includes some variables not considered by any of the others. Also, Stone's list of local authorities includes large Metropolitan Counties such as Greater London, Greater Manchester and Merseyside as single entities, whereas these large authorities are subdivided into smaller areas in the other analyses. A comparison of the clusters obtained from several different analyses is given by Jolliffe *et al.* (1986).

Table 9.1. Demographic variables used in the analysis of 46 English counties.

1. Population density—numbers per hectare
2. Percentage of population aged under 16
3. Percentage of population above retirement age
4. Percentage of men aged 16–65 who are employed
5. Percentage of men aged 16–65 who are unemployed
6. Percentage of population owning their own home
7. Percentage of households which are 'overcrowded'
8. Percentage of employed men working in industry
9. Percentage of employed men working in agriculture
10. (Length of public roads)/(area of county)
11. (Industrial floor space)/(area of county)
12. (Shops and restaurant floor space)/(area of county)

Table 9.2. Coefficients for first four PCs: English counties data.

Component number		1	2	3	4
Variable	1	0.35	−0.19	0.29	0.06
	2	0.02	0.60	−0.03	0.22
	3	−0.11	−0.52	−0.27	−0.36
	4	−0.30	0.07	0.59	−0.03
	5	0.31	0.05	−0.57	0.07
	6	−0.29	0.09	−0.07	−0.59
	7	0.38	0.04	0.09	0.08
	8	0.13	0.50	−0.14	−0.34
	9	−0.25	−0.17	−0.28	0.51
	10	0.37	−0.09	0.09	−0.18
	11	0.34	0.02	−0.00	−0.24
	12	0.35	−0.20	0.24	0.07
Eigenvalue		6.27	2.53	1.16	0.96
Cumulative percentage of total variation		52.3	73.3	83.0	90.9

As in other analyses of local authorities, PCA is used in Stone's analysis in two ways; first, to summarize and explain the major sources of variation in the data, and second, to provide a visual display on which to judge the adequacy of the clustering.

Table 9.2 gives the variances and coefficients for the first four PCs, using the correlation matrix, for Stone's data. It is seen that the first two components account for 73% of the total variation, but that most of the rules of Section 6.1 would retain four variables (the fifth eigenvalue is 0.41).

There are fairly clear interpretations for each of the first three PCs. The first PC provides a contrast between urban and rural areas, with positive coefficients for variables which are high in urban areas, such as densities of population, roads, and industrial and retail floor space; negative coefficients occur for owner occupation, percentage of employed men in agriculture, and overall employment level, which tend to be higher in rural areas. The main contrast for component 2 is between the percentages of the population below school-leaving age and above retirement age. This component is therefore a measure of the age of the population in each county, and it identifies, at one extreme, the south coast retirement areas.

The third PC contrasts employment and unemployment rates. This contrast is also present in the first Urban versus Rural PC, so that the third PC is measuring variation in employment/unemployment rates within rural areas and within urban areas, rather than between the two types of area.

Turning now to the cluster analysis of the data, Stone (1984) examines several different clustering methods, and also considers the analysis with and without Greater London, which is very different from any other area, but

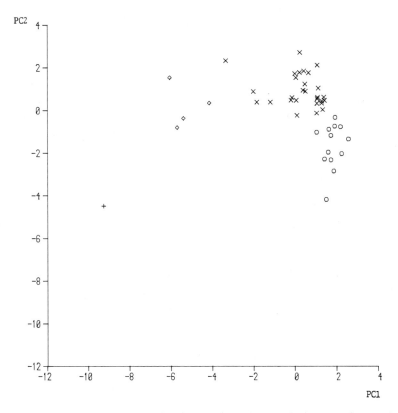

Figure 9.4. English counties: complete-linkage four-cluster solution superimposed on a plot of the first two PCs.

which makes surprisingly little difference to the analysis. Figure 9.4 shows the position of the 46 counties with respect to the first two PCs, with the four-cluster solution obtained using complete-linkage cluster analysis (see Everitt, 1980, p. 11) indicated by different symbols for different clusters. The results for complete-linkage are fairly similar to those found by several of the other clustering methods which were investigated.

In the four-cluster solution, the single observation at the bottom-left of the diagram is Greater London and the four-county cluster at the top-left consists of other metropolitan counties. The counties at the right of the diagram are more rural, confirming the interpretation of the first PC given earlier. The split between the larger groups at the right of the plot is rather more arbitrary, but as might be expected from the interpretation of the second PC, most of the retirement areas have similar values in the vertical direction; they are all in the bottom part of the diagram. Conversely, many of the counties towards the top have substantial urban areas within them, and so have somewhat lower values on the first PC as well.

The clusters are rather nicely located in different areas of the figure, although the separation between them is not particularly clear-cut, except for Greater London. This behaviour is fairly typical of what occurs for other clustering methods in this example, and for different numbers of clusters. For example, in the eight-cluster solution for complete-linkage clustering, one observation splits off from each of the clusters in the top-left and bottom-right parts of the diagram to form single-county clusters. The large 27-county cluster in the top-right of the plot splits into three groups containing 13, 10 and 4 counties, with some overlap between them.

This example is typical of many in which cluster analysis has been used for dissection. Examples like that of Jeffers' (1967) aphids, where a very clear-cut, and previously unknown, cluster structure is uncovered, are relatively unusual, although another illustration is given by Blackith and Reyment (1971, p. 155). In their example, a plot of the observations with respect to the second and third (out of seven) PCs shows a very clear separation into two groups. It is highly probable that in some circumstances 'projection-pursuit' methods (Tukey and Tukey, 1981) will provide a better two-dimensional space in which to view the results of a cluster analysis, than that defined by the first two PCs. However, if dissection, rather than discovery of a clear-cut cluster structure, is the objective of a cluster analysis, then there is likely to be little improvement over a plot with respect to the first two PCs.

9.3. Canonical Correlation Analysis

Suppose that $\mathbf{x}_{p_1}$, $\mathbf{x}_{p_2}$ are vectors of random variables with p_1, p_2 elements respectively. The objective of canonical correlation analysis is to find, successively, for $k = 1, 2, \ldots, \text{Min}[p_1, p_2]$, pairs $\{\mathbf{a}'_{k1}\mathbf{x}_{p_1}, \mathbf{a}'_{k2}\mathbf{x}_{p_2}\}$ of linear functions of $\mathbf{x}_{p_1}$, $\mathbf{x}_{p_2}$ respectively, called *canonical variates*, such that the correlation between $\mathbf{a}'_{k1}\mathbf{x}_{p_1}$ and $\mathbf{a}'_{k2}\mathbf{x}_{p_2}$ is maximized, subject to $\mathbf{a}'_{k1}\mathbf{x}_{p_1}$, $\mathbf{a}'_{k2}\mathbf{x}_{p_2}$ both being uncorrelated with $\mathbf{a}'_{jh}\mathbf{x}_{p_h}$, $j = 1, 2, \ldots, k - 1$; $h = 1, 2$. The name of the technique is confusingly similar to 'canonical analysis' which is used in discrimination (see Section 9.1). In fact, there is a link between the two techniques (see, for example, Mardia *et al.* 1979, Exercise 11.5.4), but this will not be discussed here.

Muller (1982) has suggested that there are advantages in calculating PCs $\mathbf{z}_{p_1}$, $\mathbf{z}_{p_2}$ separately for $\mathbf{x}_{p_1}$, $\mathbf{x}_{p_2}$ and then performing the canonical correlation analysis on $\mathbf{z}_{p_1}$, $\mathbf{z}_{p_2}$ rather than $\mathbf{x}_{p_1}$, $\mathbf{x}_{p_2}$. If $\mathbf{z}_{p_1}$, $\mathbf{z}_{p_2}$ consist of all p_1, p_2 PCs respectively then the results using $\mathbf{z}_{p_1}$, $\mathbf{z}_{p_2}$ will be equivalent to those for $\mathbf{x}_{p_1}$, $\mathbf{x}_{p_2}$. This follows since we are looking for 'optimal' linear functions of $\mathbf{z}_{p_1}$, $\mathbf{z}_{p_2}$. But $\mathbf{z}_{p_1}$, $\mathbf{z}_{p_2}$ are themselves exact linear functions of $\mathbf{x}_{p_1}$, $\mathbf{x}_{p_2}$ respectively, and, conversely, $\mathbf{x}_{p_1}$, $\mathbf{x}_{p_2}$ are exact linear functions of $\mathbf{z}_{p_1}$, $\mathbf{z}_{p_2}$ respectively, so that we are equivalently searching for 'optimal' linear functions of $\mathbf{x}_{p_1}$, $\mathbf{x}_{p_2}$, i.e. we have the same analysis as that based on $\mathbf{x}_{p_1}$, $\mathbf{x}_{p_2}$. However,

Muller (1982) argues that using z_{p_1}, z_{p_2} instead of x_{p_1}, x_{p_2} can make some of the theory behind canonical correlation analysis easier to understand, and it can also help in interpreting the results of such an analysis. He also illustrates the use of PCA as a dimension-reducing technique, by performing canonical correlation analysis based on just the first few elements of z_{p_1} and z_{p_2}. This works well in the example given in his paper, but cannot be expected to do so in general, for reasons similar to those already discussed in the contexts of regression (Chapter 8) and discriminant analysis (Section 9.1). There is simply no reason why those linear functions of x_{p_1} which are highly correlated with linear functions of x_{p_2} should necessarily be in the subspace spanned by the first few PCs of x_{p_1}; they could equally well be related to the last few PCs of x_{p_1}. The fact that a linear function of x_{p_1} has a small variance (as do the last few PCs) in no way prevents it from having a high correlation with some linear function of x_{p_2}.

As well as suggesting the use of PCs in canonical correlation analysis, Muller (1982) describes the closely related topic of using canonical correlation analysis to compare sets of PCs. This will be discussed further in Section 11.5.

9.3.1. Example

Jeffers (1978, p. 136) considered an example with 15 variables measured on 272 sand and mud samples, taken from various locations in Morecambe Bay, off the north-west coast of England. The variables were of two types: eight variables were chemical or physical properties of the sand or mud samples, and seven variables measured the abundance of seven groups of invertebrate species in the samples. The relationships between the two groups of variables, describing environment and species, were of interest, so that canonical correlation analysis was an obvious technique to use.

Table 9.3 gives the coefficients for the first two pairs of canonical variates, together with the correlation between each pair (the *canonical correlations*). The definitions of each variable are not given here—see Jeffers (1978, pp. 103, 107). The first canonical variate for species is dominated by a single species, but the corresponding canonical variate for the environmental variables involves non-trivial coefficients for four of the variables, though, in fact, it is not too difficult to interpret (Jeffers, 1978, p. 138). The second pair of canonical variates has fairly large coefficients for three species and three environmental variables.

Jeffers (1978, pp. 105–109) also looks at PCs for the environmental and species variables separately, and concludes that four and five PCs, respectively, are necessary to account for most of the variation in each group. He goes on to look, informally, at the between-group correlations for each set of retained PCs.

Instead of simply looking at the individual correlations between PCs for

Table 9.3. Coefficients for the first two canonical variates in a canonical correlation analysis of species and environmental variables.

		First canonical variates	Second canonical variates
Environmental variables	x_1	0.03	0.17
	x_2	0.51	0.52
	x_3	0.56	0.49
	x_4	0.37	0.67
	x_5	0.01	−0.08
	x_6	0.03	0.07
	x_7	−0.00	0.04
	x_8	0.53	−0.02
Species variables	x_9	0.97	−0.19
	x_{10}	−0.06	−0.25
	x_{11}	0.01	−0.28
	x_{12}	0.14	0.58
	x_{13}	0.19	0.00
	x_{14}	0.06	0.46
	x_{15}	0.01	0.53
Canonical correlation		0.559	0.334

different groups, an alternative would be to do a canonical correlation analysis based only on the retained PCs, as suggested by Muller (1982). In the present example, this analysis gives values of 0.420 and 0.258 for the first two canonical correlations, compared with 0.559 and 0.338 when all the variables are used. The first two canonical variates for the environmental variables, and the first canonical variate for the species variables, are each dominated by a single PC, and the second canonical variate for the species variables has two non-trivial coefficients. Thus, the canonical variates for PCs look, at first sight, easier to interpret than those based on the original variables. However, it must be remembered that, even if only one PC occurs in a canonical variate, the PC itself is not necessarily an easily interpreted entity. For example, the environmental PC which dominates the first canonical variate for the environmental variables has six large coefficients. Furthermore, the between-group relationships found by canonical correlation analysis of PCs are different, in this example, from those found from canonical correlation analysis on the original variables.

CHAPTER 10

Outlier Detection, Influential Observations and Robust Estimation of Principal Components

This chapter deals with three related topics, which are all concerned with situations where some of the observations may, in some way, be atypical of the bulk of the data.

First, we discuss the problem of detecting *outliers* in a set of data. Outliers are generally viewed as observations which are a long way from, or inconsistent with, the remainder of the data. Such observations can, but need not, have a drastic and disproportionate effect on the results of various analyses of a data set. Numerous methods have been suggested for detecting outliers (see Barnett and Lewis, 1978; Hawkins, 1980); some of the methods use PCs, and these methods are described in Section 10.1.

The techniques described in Section 10.1 are useful regardless of the type of statistical analysis to be performed, but, in Sections 10.2 and 10.3, we look specifically at the case where a PCA is being done. Depending on their position, outlying observations may or may not have a large effect on the results of the analysis. It is of interest to determine which observations do indeed have a large effect. Such observations are called *influential observations* and are discussed in Section 10.2.

Given that certain observations are outliers or influential, it may be desirable to adapt the analysis to remove or diminish the effects of such observations, i.e. the analysis is made *robust*. Robust analyses in many branches of statistics have been a subject of much research effort in recent years (see, for example, Huber (1981) for some of the theoretical background, and Hoaglin *et al.* (1983) for a more readable approach), and robustness with respect to distributional assumptions, as well as with respect to outlying or influential observations, may be of interest. Because PCA does not, for most purposes, need distributional assumptions (see Section 3.7), robustness with respect to distribution is generally less important in PCA than robustness with respect to outlying or influential observations. A number of techniques have been suggested for robustly estimating PCs, and these are discussed in the third section of this chapter; the final section presents a few concluding remarks.

10.1. Detection of Outliers Using Principal Components

There is no formal, widely accepted, definition of what is meant by an 'outlier'. The books on the subject by Barnett and Lewis (1978) and Hawkins (1980) both rely on informal, intuitive definitions, namely that outliers are observations which are in some way different from, or inconsistent with, the remainder of a data set. For p-variate data, this definition implies that outliers are a long way from the rest of the observations in the p-dimensional space defined by the variables. Numerous procedures have been suggested for detecting outliers with respect to a single variable, and many of these are reviewed by Barnett and Lewis (1978) and Hawkins (1980). The literature on multivariate outliers is less extensive, with each of the above books containing only one chapter (comprising less than 10% of their total content) on the subject. Several approaches to the detection of multivariate outliers use PCs, and these will now be discussed in some detail. As well as the methods described in this section, which use PCs in fairly direct ways to identify potential outliers, techniques for robustly estimating PCs (see Section 10.3) may also be used to detect outlying observations.

A major problem in detecting multivariate outliers is that an observation may not be extreme on any of the original variables, but it can still be an outlier because it does not conform with the correlation structure of the remainder of the data. It is impossible to detect such outliers by looking solely at the original variables one at a time. As a simple example, suppose that heights and weights are measured for a sample of healthy children of various ages between 5 and 15. Then an 'observation' with height and weight of 70 in. and 60 lb respectively will not be particularly extreme on either the height or weight variables individually, since 70 in. is a plausible height for the older children, and 60 lb is a plausible weight for the youngest children. However, the combination (70 in., 60 lb) is virtually impossible, and will be a clear outlier because it combines a large height with a small weight, thus violating the general pattern of a positive correlation between the two variables. Such an outlier would be apparent on a plot of the two variables (see Figure 10.1), but, if the number of variables (p) is large, it is quite possible that outliers will not be apparent on any of the $\frac{1}{2}p(p-1)$ plots of two variables at a time. Thus, for large p, we need to consider the possibility that outliers will manifest themselves in directions other than those which are detectable from simple plots of pairs of the original variables.

Outliers can be of many types, which complicates any search for directions

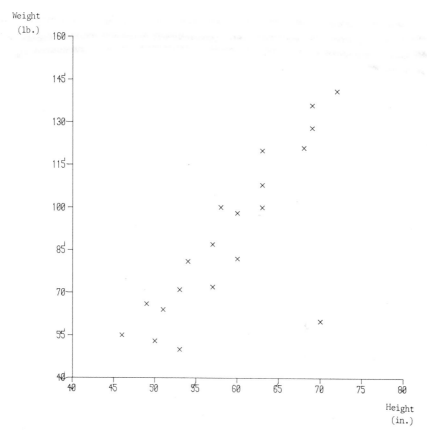

Figure 10.1. Example of an outlier which is not detectable by looking at one variable at a time.

in which outliers occur. However, there are good reasons for looking at the directions defined by either the first few, or the last few, PCs in order to detect outliers. The first few, and last few, PCs will detect different types of outlier, and, in general, the last few are more likely to provide additional information which is not available in plots of the original variables.

As discussed in Gnanadesikan and Kettenring (1972), the outliers which are detectable from a plot of the first few PCs are those which *inflate* variances and covariances. But if an outlier is the cause of a large increase in one or more of the variances of the original variables, then it will be extreme on those variables, and thus detectable by looking at plots of the original variables. Similarly, an observation which inflates a covariance (or correlation) between two variables will generally be clearly visible on a plot of these two variables, and will often be extreme with respect to one or both of these variables looked at individually.

By contrast, the last few PCs may detect outliers which are not apparent with respect to the original variables. A strong correlation structure between variables will imply that there are linear functions of the variables with small variances (compared to the variances of the original variables). In the simple height-and-weight example described above, height and weight have a strong positive correlation, so it is possible to write

$$x_2 = \beta x_1 + \varepsilon,$$

where x_1, x_2 are height and weight measured about their sample means, β is a positive constant, and ε is a random variable with a much smaller variance than x_1 or x_2. Therefore the linear function

$$x_2 - \beta x_1$$

has a small variance, and the second (i.e. last) PC in an analysis of x_1, x_2 will have a similar form, namely $a_{22}x_2 - a_{12}x_1$, where $a_{12}, a_{22} > 0$. Calculation of the value of this second PC for each observation will detect observations, such as (70 in., 60 lb), which are outliers with respect to the correlation structure of the data, though not necessarily with respect to individual variables. Figure 10.2 shows a plot of the data from Figure 10.1, with respect to the PCs derived from the correlation matrix. The outlying observation is 'average' for the first PC, but very extreme for the second.

This argument generalizes readily when the number of variables p is greater than two; by examining the values of the last few PCs, we may be able to detect observations which violate the correlation structure imposed by the bulk of the data, but which are not necessarily aberrant with respect to individual variables. Of course, it is possible that, if the sample size is relatively small, or if a few observations are sufficiently different from the rest, then the outlier(s) may so strongly influence the last few PCs, that these PCs now reflect mainly the position of the outlier(s) rather than the structure of the majority of the data. One way of avoiding this masking or camouflage of outliers is to compute PCs leaving out one (or more) observations, and then calculate, for the deleted observations, the values of the last PCs based on the reduced data set. To do this for each observation is a heavy computational burden, but it might be worthwhile in small samples where such camouflaging is, in any case, more likely to occur. Alternatively, if PCs are estimated robustly (see Section 10.3), then the influence of outliers on the last few PCs should be reduced, and it may be unnecessary to repeat the analysis with each observation deleted.

As well as simple plots of observations with respect to PCs, it is possible to set up more formal tests for outliers, based on PCs, assuming that the PCs are normally distributed. Strictly, this assumes that $\mathbf{x}$ has a multivariate normal distribution, but, in fact, because the PCs are linear functions of p random variables, an appeal to the Central Limit Theorem may justify approximate normality for the PCs even when the original variables are not normal. A battery of tests is then available for each individual PC, namely

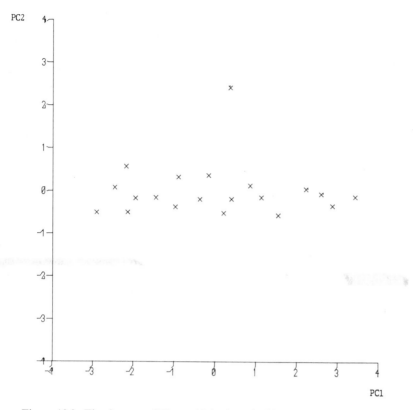

Figure 10.2. The data set of Figure 10.1, plotted with respect to its PCs.

those for testing for the presence of outliers in a sample of (univariate) normal data—see Hawkins (1980, Chapter 3) and Barnett and Lewis (1978, Section 3.4.3, which describes 44 tests!)

Other tests, which combine information from several PCs, rather than examining one at a time, are described by Gnanadesikan and Kettenring (1972) and Hawkins (1974), and some of these will now be discussed. In particular, we shall define four statistics which will be denoted $d_{1i}^2, d_{2i}^2, d_{3i}^2$ and d_{4i}.

The last few PCs are likely to be more useful than the first few in detecting outliers which are not apparent from the original variables, so one possible test statistic, d_{1i}^2, suggested by Rao (1964), and discussed further by Gnanadesikan and Kettenring (1972), is the sum of squares of the values of the last q ($< p$) PCs, i.e.

$$d_{1i}^2 = \sum_{k=p-q+1}^{p} z_{ik}^2, \qquad (10.1.1)$$

where z_{ik} is the value of the kth PC for the ith observation. $d_{1i}^2, i = 1, 2, \ldots, n$

should, approximately, be independent observations from a gamma distribution if there are no outliers, so that a gamma probability plot with suitably estimated shape parameter may expose outliers (Gnanadesikan and Kettenring, 1972). Note that d_{1i}^2, computed separately for several populations, is also used in a form of discriminant analysis (SIMCA) by Wold *et al.* (1983)— see Section 9.1.

A possible criticism of the statistic d_{1i}^2 is that it still gives insufficient weight to the last few PCs, especially if q, the number of PCs contributing to d_{1i}^2, is close to p. Because the PCs have decreasing variance with increasing index, the values of z_{ik}^2 will typically become smaller as k increases, and d_{1i}^2 therefore implicitly gives the PCs decreasing weight as k increases. This effect can be severe if some of the PCs have very small variances, and this is unsatisfactory, since it is precisely the low-variance PCs which may be most effective in determining the presence of certain types of outlier.

An alternative is to give the components equal weight and this can be achieved by replacing z_{ik} by $z_{ik}^* = z_{ik}/l_k^{1/2}$, where l_k is the variance of the kth sample PC. In this case the sample variances of the z_{ik}^*'s will all be equal to unity. Hawkins (1980, Section 8.2) justifies this particular renormalization of the PCs by noting that the renormalized PCs, in reverse order, are the uncorrelated linear functions of $\mathbf{x}$, $\tilde{\mathbf{a}}_p'\mathbf{x}, \tilde{\mathbf{a}}_{p-1}'\mathbf{x}, \ldots, \tilde{\mathbf{a}}_1'\mathbf{x}$, which, when constrained to have unit variances, have coefficients $\tilde{a}_{jk}$ which sucessively maximize the criterion $\sum_{j=1}^p \tilde{a}_{jk}^2$, for $k = p, p-1, \ldots, 1$. Maximization of this criterion is desirable because, given the fixed-variance property, linear functions which have large absolute values for their coefficients will be more sensitive to outliers than those with small coefficients. It should be noted that when $q = p$, the statistic

$$d_{2i}^2 = \sum_{k=p-q+1}^{p} z_{ik}^2/l_k \tag{10.1.2}$$

becomes $\sum_{k=1}^p z_{ik}^2/l_k$, which is simply the Mahalanobis distance between the ith observation and the sample mean, which is defined as $(\mathbf{x}_i - \bar{\mathbf{x}})'\mathbf{S}^{-1}(\mathbf{x}_i - \bar{\mathbf{x}})$. This follows because $\mathbf{S} = \mathbf{AL}^2\mathbf{A}'$ where, as usual, $\mathbf{L}^2$ is the diagonal matrix whose kth diagonal element is l_k, and $\mathbf{A}$ is the matrix whose (j, k)th element is a_{jk}. Furthermore,

$$\mathbf{S}^{-1} = \mathbf{AL}^{-2}\mathbf{A}',$$

$$\mathbf{x}_i' = \mathbf{z}_i'\mathbf{A}',$$

$$\bar{\mathbf{x}}' = \bar{\mathbf{z}}'\mathbf{A}',$$

and so

$$(\mathbf{x}_i - \bar{\mathbf{x}})'\mathbf{S}^{-1}(\mathbf{x}_i - \bar{\mathbf{x}}) = (\mathbf{z}_i - \bar{\mathbf{z}})'\mathbf{A}'\mathbf{AL}^{-2}\mathbf{A}'\mathbf{A}(\mathbf{z}_i - \bar{\mathbf{z}})$$

$$= (\mathbf{z}_i - \bar{\mathbf{z}})'\mathbf{L}^{-2}(\mathbf{z}_i - \bar{\mathbf{z}})$$

$$= \sum_{k=1}^{p} z_{ik}^2/l_k,$$

where z_{ik} is the kth PC score for the ith observation, measured about the mean of the scores for all observations.

Gnanadesikan and Kettenring (1972) also consider the statistic

$$d_{3i}^2 = \sum_{k=p-q+1}^{p} l_k z_{ik}^2, \qquad (10.1.3)$$

with $q = p$, which emphasizes observations which have a large effect on the *first few* PCs, and is equivalent to $(\mathbf{x}_i - \bar{\mathbf{x}})'\mathbf{S}(\mathbf{x}_i - \bar{\mathbf{x}})$. As stated earlier, the first few PCs are useful in detecting some types of outlier, and d_{3i}^2 will emphasize such outliers. However, we repeat that such outliers will often be detectable from plots of the original variables, unlike the outliers exposed by the last few PCs. Various types of outlier, including some which are extreme with respect to both the first few *and* and the last few PCs, are illustrated in the examples given later in this section.

Hawkins (1974) prefers to use d_{2i}^2 with $q < p$, rather than $q = p$ (again, in order to emphasize the low-variance PCs), and he considers how to choose an appropriate value for q. This is a rather different problem from that considered in Section 6.1, since we now wish to decide how many of the PCs, *starting with the last*, rather than starting with the first, need to be retained. Hawkins (1974) suggests three possibilities for choosing q, including the 'opposite' of Kaiser's rule (Section 6.1.2)—i.e. retain PCs with eigenvalues less than unity. In an example he selects q as a compromise between values suggested by his three rules.

Hawkins (1974) also shows that outliers can be successfully detected using the statistic,

$$d_{4i} = \max_{p-q+1 \le k \le p} |z_{ik}^*|, \qquad (10.1.4)$$

and similar methods for choosing q are again suggested. Fellegi (1975), too, is enthusiastic about the performance of the statistic d_{4i}. Hawkins and Fatti (1984) claim that outlier detection is improved still further if the last q re-normalized PCs are rotated using varimax rotation (see Chapter 7) before computing d_{4i}. The test statistic for the ith observation then becomes the maximum absolute value of the last q renormalized and rotated PCs, evaluated for that observation.

The exact distributions for $d_{1i}^2, d_{2i}^2, d_{3i}^2$ and d_{4i} can easily be deduced if we assume that the observations are from a multivariate normal distribution with mean $\boldsymbol{\mu}$, and covariance matrix $\boldsymbol{\Sigma}$, where $\boldsymbol{\mu}, \boldsymbol{\Sigma}$ are both known (see Hawkins (1980, p. 113) for the results for d_{2i}^2, d_{4i}).

Both d_{3i}^2, and d_{2i}^2 when $q = p$, as well as d_{1i}^2, will have (approximate) gamma distributions if no outliers are present and if normality can be (approximately) assumed (Gnanadesikan and Kettenring, 1972), so that gamma probability plots of d_{2i}^2 (with $q = p$) and d_{3i}^2 can again be used to look for outliers. However, in practice $\boldsymbol{\mu}, \boldsymbol{\Sigma}$ are unknown, and the data will often not have a multivariate normal distribution, so that any distributional results derived under the restrictive assumptions can only be approximations. In order to be satisfactory, such approximations need not be particularly accurate, since outlier detection is usually concerned with finding observations which are blatantly different from the rest, corresponding to very small significance levels for the test statistics. An observation which is 'barely significant at 5%' will typically not be of interest, so that there is no great incentive to compute significance levels very accurately. The outliers which we wish to detect should 'stick out like a sore thumb' provided we find the right direction in which to view the data; the problem in multivariate outlier detection is to find appropriate directions.

The discussion in this section has been in general terms, i.e. PCs can be used to detect outliers in any multivariate data set, regardless of the subsequent analysis which is envisaged for that data set. This includes the case where the data are collected in order to perform a multiple regression, or more generally a multivariate regression. For multiple regression, Hocking (1984) suggests that plots of PCs derived from $(p + 1)$ variables, consisting of p predictor variables together with the dependent variable (as used in latent root regression—see Section 8.4) tend to reveal outliers, together with observations which are highly influential (see the next section of this chapter) for the regression equation. Plots of PCs derived from the predictor variables only will also tend to reveal influential observations. Hocking's (1984) suggestions are illustrated with an example, but no indication is given of whether the first few, or last few, PCs are more likely to be useful—his example has only three predictor variables, so it is easy to look at all possible plots. In the case of multivariate regression, another possibility for detecting outliers (Gnanadesikan and Kettenring, 1972) is to look at the PCs of the (multivariate) residuals from the regression analysis.

Another specialized field in which the use of PCs has been proposed in order to detect outlying observations, is that of quality control. Jackson and Mudholkar (1979) suggest the statistic d_{1i}^2, and examine an approximation to its distribution. They prefer d_{1i}^2 to d_{2i}^2 for computational reasons, and because of its intuitive appeal as a sum of squared residuals from the $(p - q)$-dimensional space defined by the first $(p - q)$ PCs. However, Jackson and Hearne (1979) indicate that the 'opposite' of d_{2i}^2, in which the sum of squares of the first few, rather than last few, renormalized PCs is calculated, may be useful in quality control, when the objective is to look for *groups* of 'out-of-control', or outlying, observations, rather than single outliers. Their basic statistic is decomposed to give separate information about variation *within* the sample of potentially outlying observations, and about the difference between the sample mean and some known standard value. In addition, they propose an alternative statistic based on absolute, rather than squared, values of PCs. Jackson and Mudholkar (1979) also extended their proposed control procedure, based on d_{1i}^2, to the multiple-outlier case. Coleman (1985) suggests that when using PCs in quality control, the PCs should be estimated robustly—see Section 10.3.

A different way of using PCs to detect outliers was proposed by Gabriel and Zamir (1979). This proposal uses the idea of weighted PCs, and will be discussed further in Section 12.1.

Before turning to examples, note that an example in which outliers are detected using PCs, in a rather different way, has already been given in Section 5.6. In that example, Andrews' curves (Andrews, 1972) were computed using PCs and some of the observations stood out as different from the others when plotted as curves. Further examination of these different observations showed that they were indeed 'outlying' in some respects, compared to the remaining observations.

10.1.1. Examples

In this section, one example will be discussed in some detail, while two others will be described more briefly.

Anatomical Measurements

A set of seven anatomical measurements on 28 students was discussed in Section 5.1.1 and it was found that on a plot of the first two PCs (Figure 5.1) there was an extreme observation on the second PC. If the measurements of this individual are examined in detail, it is found that he has an anomalously small head circumference. Whereas the other 27 students all had head girths in the narrow range 21–24 cm, this student (No. 16) had a measurement of 19 cm. It was impossible to check whether this was an incorrect measurement or whether student 16 had an unusually small head (his other measurements were close to average), but it is clear that this observation would be regarded as an 'outlier' according to most definitions of the term.

This particular outlier was detected on the second PC, and it was suggested above that any outliers detected by high-variance PCs will usually be detectable on examination of individual variables; this is indeed the case here. Another point concerning this observation is that it is so extreme on the second PC that it may be suspected that it alone is largely responsible for the direction of this PC. This question will be investigated at the end of the next section, which deals with influential observations.

Figure 5.1 indicates one other possible outlier, at the extreme left of the diagram. This turns out to be the largest student in the class—190 cm (6′ 3″) tall, with all measurements except head girth at least as large as all other 27 students. There is no suspicion here of any incorrect measurements.

Turning now to the last few PCs, we would hope to detect any observations which are 'outliers' with respect to the correlation structure of the data. Figure 10.3 gives a plot of the observations with respect to the last two PCs, and Table 10.1 gives the values of d_{1i}^2, d_{2i}^2 and d_{4i}, defined in equations (10.1.1), (10.1.2) and (10.1.4), respectively, for the six 'most extreme' observations on each statistic, where the number of PCs included, q, is 1, 2 or 3. The

Table 10.1. Anatomical measurements: values of d_{1i}^2, d_{2i}^2, d_{4i} for the most extreme observations.

| | | | | | | | | Number of PCs used, q | | | | | | |
| $q = 1$ | | $q = 2$ | | | | | | | $q = 3$ | | | | | |
d_{1i}^2	Obs. No.	d_{1i}^2	Obs. No.	d_{2i}^2	Obs. No.	d_{4i}	Obs. No.	d_{1i}^2	Obs. No.	d_{2i}^2	Obs. No.	d_{4i}	Obs. No.
0.81	15	1.00	7	7.71	15	2.64	15	1.55	20	9.03	20	2.64	15
0.47	1	0.96	11	7.69	7	2.59	11	1.37	5	7.82	15	2.59	5
0.44	7	0.91	15	6.70	11	2.01	1	1.06	11	7.70	5	2.59	11
0.16	16	0.48	1	4.11	1	1.97	7	1.00	7	7.69	7	2.53	20
0.15	4	0.48	23	3.52	23	1.58	23	0.96	1	7.23	11	2.01	1
0.14	2	0.36	12	2.62	12	1.49	27	0.93	15	6.71	1	1.97	7

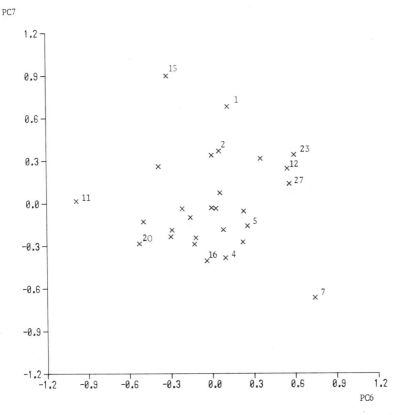

Figure 10.3. Anatomical measurements: plot of observations with respect to the last two PCs.

observations which correspond to the most extreme values of d_{1i}^2, d_{2i}^2 and d_{4i} are identified in Table 10.1, and also on Figure 10.3.

Note that when $q = 1$, the observations have the same ordering for all three statistics, so only the values of d_{1i}^2 are given in Table 10.1. When q is increased to 2 or 3, the six most extreme observations are the same (in a slightly different order) for both d_{1i}^2 and d_{2i}^2. With the exception of the sixth most extreme observation for $q = 2$, the same observations are also identified by d_{4i}. Although the sets of the six most extreme observations are virtually the same for d_{1i}^2, d_{2i}^2 and d_{4i}, there are some differences in ordering. The most notable example is observation 15, which, for $q = 3$, is most extreme for d_{4i}, but only sixth most extreme for d_{1i}^2.

Observations 1, 7 and 15 are extreme on all seven statistics given in Table 10.1, due to large contributions from the final PC alone for observation 15, the last two PCs for observation 7, and the fifth and seventh PCs for observation 1. Observations 11 and 20, which are not extreme for the final PC, appear in the columns for $q = 2$ and 3 because of extreme behaviour on the sixth PC for observation 11, and on both the fifth and sixth PCs for observation 20. Observation 16, which was discussed earlier as a clear outlier on the second PC, appears in the list for $q = 1$, but is not notably extreme for any of the last three PCs.

Most of the observations identified in Table 10.1 are near the edge of the plot given in Figure 10.3, although observations 2, 4, 5, 12, 16, 20, 23 and 27 are close to the main body of the data. However, observations 7, 11, 15, and to a lesser extent 1, are sufficiently far from the remaining data to be worthy of further consideration. To roughly judge their 'significance', recall that, if no outliers are present and the data are approximately multivariate normal, then the values of d_{4i} are (approximately) absolute values of a normal random variable with zero mean and unit variance. The quantities given in any column of Table 10.1 are therefore the six largest among $28q$ such variables, and none of them look particularly extreme. Nevertheless, it is of interest to investigate the reasons for the outlying positions of some of the observations, and to do so it is necessary to examine the coefficients of the last few PCs. The final PC, accounting for only 1.7% of the total variation, is largely a contrast between chest and hand measurements, with positive coefficients 0.55, 0.51, and waist and height measurements which have negative coefficients -0.55, -0.32. Looking at observation 15, we find that this (male) student has the equal largest chest measurement, but that only three of the other 16 male students are shorter than him, and only two have a smaller waist measurement. Similar analyses can be done for other observations in Table 10.1. For example, observation 20 is extreme on the fifth PC. This PC, which accounts for 2.7% of the total variation, is mainly a contrast between height and forearm length, with coefficients 0.67, -0.52 respectively. Observation 20 is (jointly with one other) the shortest student of the 28, but only one of the other ten women has a larger forearm measurement. Thus, observations 15 and 20, and other observations indicated as extreme by the last few PCs, are students for whom some aspects of their physical measurements contradict the general positive correlation between all seven measurements.

Household Formation Data

These data were described in Section 8.7.2, and are discussed in detail by Garnham (1979) and Bassett et al. (1980). Section 8.7.2 gave the results of a PC regression of average annual total income per adult on 28 other demographic variables, for 168 local government areas in England and Wales. Garnham (1979) also examined plots of the last few and first few PCs of the 28 predictor variables, in an attempt to detect outliers. Two such plots, for the first two, and last two, PCs are reproduced in Figures 10.4 and 10.5. An interesting aspect of these figures is that the most extreme observations with

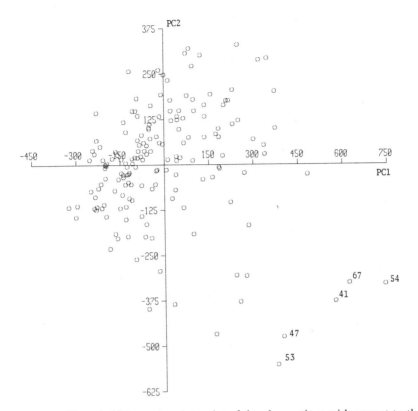

Figure 10.4. Household formation data: plot of the observations with respect to the first two PCs.

respect to the last two PCs, namely observations 54, 67, 41 (and 47 and 53) are also among the most extreme with respect to the first two PCs. Some of these observations are, in addition, in outlying positions on plots of other low-variance PCs. The most blatant case is observation 54 which is among the few most extreme observations on PCs 24 to 28 inclusive, as well as on PC1. This observation is 'Kensington and Chelsea', which must be an outlier with respect to several variables individually, as well as differing in correlation structure from most of the remaining observations.

In addition to plotting the data with respect to the last few and first few PCs, Garnham (1979) examined the statistics d_{1i}^2 for $q = 1, 2, \ldots, 8$, using gamma plots, and also looked at normal probability plots of the values of various PCs. As a combined result of these analyses, he identified six likely outliers, the five mentioned above together with observation 126, which was moderately extreme according to several analyses.

The PC regression was then repeated without these six observations, and

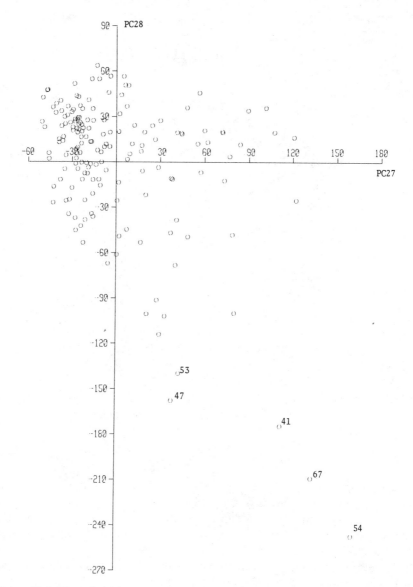

Figure 10.5. Household formation data: plot of the observations with respect to the last two PCs.

the results of the regression were noticeably changed, and were better in two respects than those derived from all the observations. The number of PCs which it was necessary to retain in the regression was decreased, and the prediction accuracy was improved, with the standard error of prediction reduced to 77.3% of that for the full data set.

Trace Element Concentrations

The final example of this section is discussed by Hawkins and Fatti (1984). The data consist of measurements of the log-concentrations of 12 trace elements in 75 rock-chip samples. In order to detect outliers, Hawkins and Fatti simply look at the values for each observation on each variable, on each PC, and on PCs rotated by varimax rotation. To decide whether an observation is an outlier, a cut-off is defined assuming normality, and using a Bonferroni bound with significance level 0.01. On the original variables, only two observations satisfy this criterion for outliers, but the number of outliers increases to seven if PCs are used. Six of these seven outlying observations are extreme on one of the last four PCs, and each of these low-variance PCs accounts for less than 1% of the total variation. The PCs are thus again detecting observations whose correlation structure differs from the bulk of the data, rather than those which are extreme on individual variables. Indeed, one of the 'outliers' on the original variables is not detected by the PCs.

When rotated PCs are considered, nine observations are declared to be outliers, including all those detected by the original variables and by the unrotated PCs. There is a suggestion, then, that rotation of the PCs provides an even more powerful tool for detecting outliers.

10.2. Influential Observations in a Principal Component Analysis

The idea of influence is a fairly recent one in statistics, and is a somewhat different concept from that of outliers. Outliers are generally thought of as observations which are in some way atypical of a data set but, depending on the analysis done for a data set, removal of an outlier may or may not have a substantial effect on the results of that analysis. Observations whose removal does have a large effect are called influential, and, whereas most influential observations are outliers in some respect, outliers need not be at all influential. Also, whether or not an observation is influential depends on the analysis being done on the data set; observations which are influential for one type of analysis or parameter of interest may not be so for a different analysis or parameter. This behaviour is illustrated in PCA where observations which are influential for the coefficients of a PC are not necessarily influential for the variance of that PC, and vice versa.

The intuitive definition of the influence of an observation on a statistic such as the kth eigenvalue l_k, or eigenvector $\mathbf{a}_k$, of a sample covariance matrix is simply the change in l_k (or $\mathbf{a}_k$), perhaps normalized in some way, when the observation is deleted from the sample. The problem with influence defined in this way is that there may be no closed form for the influence function, and it needs to be computed afresh for each different sample. Various other definitions of sample influence have been proposed (see, for example, Cook and Weisberg, 1982, Section 3.4); some of these have closed-form expressions for regression coefficients (Cook and Weisberg, 1984, Section 3.4), but not for the statistics of interest in PCA. Alternatively, a theoretical influence function can be defined, which can be expressed as a once-and-for-all formula, and provided sample sizes are not too small, can be used to estimate the influence of actual and potential observations within a sample.

To define the theoretical influence function, suppose that $\mathbf{y}$ is a p-variate random vector, and let $\mathbf{y}$ have cumulative distribution function (c.d.f.) $F(\mathbf{y})$. If $\boldsymbol{\theta}$ is a vector of parameters of the distribution of $\mathbf{y}$ (such as λ_k, $\boldsymbol{\alpha}_k$, the kth eigenvalue and eigenvector of the covariance matrix of $\mathbf{y}$) then $\boldsymbol{\theta}$ can be written as a function of F. Now let $F(\mathbf{y})$ be perturbed to become

$$\tilde{F}(\mathbf{y}) = (1 - \varepsilon)F(\mathbf{y}) + \varepsilon\delta_{\mathbf{x}},$$

where $0 < \varepsilon < 1$ and $\delta_{\mathbf{x}}$ is the c.d.f. of the random variable which takes the value $\mathbf{x}$ with certainty; let $\tilde{\boldsymbol{\theta}}$ be the value of $\boldsymbol{\theta}$ when F becomes $\tilde{F}$. Then the influence function $I(\mathbf{x}; \boldsymbol{\theta})$ for $\boldsymbol{\theta}$, evaluated at a value $\mathbf{x}$ of the random variable is defined to be (Hampel, 1974)

$$I(\mathbf{x}; \boldsymbol{\theta}) = \lim_{\varepsilon \to 0} (\tilde{\boldsymbol{\theta}} - \boldsymbol{\theta})/\varepsilon.$$

Alternatively, if $\tilde{\boldsymbol{\theta}}$ is expanded about $\boldsymbol{\theta}$ as a power series in ε i.e.

$$\tilde{\boldsymbol{\theta}} = \boldsymbol{\theta} + \mathbf{c}_1\varepsilon + \mathbf{c}_2\varepsilon^2 + \cdots, \qquad (10.2.1)$$

then the influence function is the coefficient, $\mathbf{c}_1$, of ε in this expansion.

Some of the above may appear somewhat abstract, but in many situations an expression can be derived for $I(\mathbf{x}; \boldsymbol{\theta})$ without too much difficulty, and, as we shall see in examples below, $I(\mathbf{x}; \boldsymbol{\theta})$ can provide valuable guidance about the influence of individual observations in *samples*.

The influence function for λ_k and $\mathbf{a}_k$ is given by Radhakrishnan and Kshirsagar (1981) and by Critchley (1985), for the case of covariance matrices. Critchley (1985) also discusses various sample versions of the influence function, and considers the coefficients of ε^2, as well as ε, in the expansion (10.2.1). Calder (1985) gives results for correlation matrices, which are somewhat different in character from those for covariance matrices. Further, as yet unpublished, work has been done on this topic by Patricia Calder at the University of Kent, but only an outline of the main results is given here.

For covariance matrices, the theoretical influence function for λ_k can be written very simply as

$$I(\mathbf{x}; \lambda_k) = z_k^2 - \lambda_k, \qquad (10.2.2)$$

where z_k is the value of the kth PC for the given value of $\mathbf{x}$ i.e. z_k is the kth element of $\mathbf{z}$, where $\mathbf{z} = \mathbf{A}'\mathbf{x}$, using the same notation as before. Thus, the influence of an observation on λ_k depends only on its position with respect to the kth component; an observation can be extreme on any or all of the other components without affecting λ_k. This illustrates the point made earlier that outlying observations need not necessarily be influential for every part of an analysis.

For correlation matrices, $I(\mathbf{x}; \lambda_k)$ takes a different form, which can be written most conveniently as

$$I(\mathbf{x}; \lambda_k) = \sum_{\substack{i=1 \\ i \neq j}}^{p} \sum_{j=1}^{p} \alpha_{ki} \alpha_{kj} I(\mathbf{x}; \rho_{ij}), \qquad (10.2.3)$$

where α_{kj} is the jth element of $\boldsymbol{\alpha}_k$, $I(\mathbf{x}; \rho_{ij}) = -\frac{1}{2}\rho_{ij}(x_i^2 + x_j^2) + x_i x_j$, and x_i, x_j are elements of $\mathbf{x}$ standardized to zero mean and unit variance. $I(\mathbf{x}; \rho_{ij})$ is the influence function for the correlation coefficient ρ_{ij}, which is given by Devlin *et al.* (1975). The expression (10.2.3) is a relatively simple one, and it shows that investigation of the influence of an observation on the correlation coefficients will be useful in determining the influence of the observation on λ_k.

There is a corresponding expression to (10.2.3) for covariance matrices, expressing $I(\mathbf{x}; \lambda_k)$ in terms of influence functions for elements of the covariance matrix. However, when (10.2.3) is written in terms of $\mathbf{x}$, or the PCs, by substituting for $I(\mathbf{x}; \rho_{ij})$, it cannot be expressed in as simple a form as (10.2.2). In particular, $I(\mathbf{x}; \lambda_k)$ now depends on $z_j, j = 1, 2, \ldots, p$, and not just on z_k. This result reflects the fact that a change to a covariance matrix may change one of the eigenvalues without affecting the others, but that this cannot happen for a correlation matrix. For a correlation matrix the sum of the eigenvalues is a constant, so that if one of them is changed, there must be compensatory changes in at least one of the others.

Expressions for $I(\mathbf{x}; \boldsymbol{\alpha}_k)$ are more complicated than those for $I(\mathbf{x}; \lambda_k)$; for example, for covariance matrices we have the expression

$$I(\mathbf{x}; \boldsymbol{\alpha}_k) = -z_k \sum_{j \neq k} z_j \boldsymbol{\alpha}_j (\lambda_j - \lambda_k)^{-1} \qquad (10.2.4)$$

compared with (10.2.2) for $I(\mathbf{x}; \lambda_k)$. A number of comments should be made concerning (10.2.4), and the corresponding expression for correlation matrices. First, and perhaps most importantly, the form of the expression is completely different from that for $I(\mathbf{x}; \lambda_k)$. It is possible for an observation to be influential for λ_k but not for $\boldsymbol{\alpha}_k$, and vice versa. This behaviour will be illustrated by an example which follows shortly.

A second related point is that, for covariance matrices, $I(\mathbf{x}; \boldsymbol{\alpha}_k)$ depends on all of the PCs, $z_1, z_2, \ldots, z_p$, unlike $I(\mathbf{x}; \lambda_k)$ which depends just on z_k. The dependence is quadratic, but involves only cross-product terms $z_j z_k, j \neq k$, and not linear or squared terms. The general shape of the influence curves, $I(\mathbf{x}; \boldsymbol{\alpha}_k)$, is hyperbolic for both covariance and correlation matrices, but the details of the function are different.

To show that theoretical influence functions are relevant to sample data, sample influence functions can be calculated by investigating the effect of adding or deleting an observation, for a given sample covariance matrix, using hypothetical samples of different sizes. These sample influence functions can then be compared with the corresponding theoretical influence functions, and there are, in fact, close relationships between the two. Alternatively, predictions from the theoretical influence function can be compared with actual changes caused by the deletion of one observation at a time from a genuine data set. The theoretical influence function typically contains unknown parameters and these must be replaced by equivalent sample quantities in such comparisons. An example is now given which illustrates that the theoretical influence function can, indeed, give a good guide to sample behaviour for moderate sample sizes.

10.2.1. Examples

Two examples are now given, both using data sets which have been discussed earlier. In the first example, we examine the usefulness of expressions for theoretical influence in predicting the actual effect of omitting observations, using the data on artistic qualities of painters described in Section 5.1.1. As a second illustration, we follow up the suggestion, made in the previous section, that an outlier was largely responsible for the form of the second PC in the student anatomical data.

Artistic Qualities of Painters

We consider again the set of four subjectively assessed variables, on 54 painters, which was described by Davenport and Studdert-Kennedy (1972) and discussed in Section 5.1.1. Tables 10.2 and 10.3 give some comparisons between the values of the influence functions obtained from expressions such as (10.2.2), (10.2.3) and (10.2.4) by substituting sample quantities l_k, a_{kj}, r_{ij} in place of the unknown λ_k, α_{kj}, ρ_{ij}, and the actual changes in eigenvalues and eigenvectors obtained when individual observations are omitted. The information given in Table 10.2 relates to PCs derived from the covariance matrix; Table 10.3 gives corresponding results for the correlation matrix. Some further explanation is necessary of exactly how the numbers in these two tables are derived.

First, the 'actual' changes in eigenvalues are precisely that—the differences between eigenvalues with and without a particular observation included in the analysis. The tables give the four largest and four smallest such changes for each PC, and identify those observations for which changes occur. The 'estimated' changes in eigenvalues given in Tables 10.2 and 10.3 are derived from multiples of (10.2.2), (10.2.3) respectively, with the value of the kth PC for each individual observation substituted for z_k, and with l_k, a_{kj}, r_{ij} replacing λ_k, α_{kj}, ρ_{ij}. The multiples are required because Change = (Influence) × ε, and we need a replacement for ε; here we have used $1/(n-1)$, where $n = 54$ is the sample size.

Table 10.2. Artistic qualities of painters: comparisons between estimated and actual influence of individual observations for the first two PCs, based on the covariance matrix.

Component number		Influence 1						Influence 2					
	Eigenvalue			Eigenvector			Eigenvalue			Eigenvector			
Estimated	Actual	Obs. No.	Estimated	Actual	Obs. No.	Estimated	Actual	Obs. No.	Estimated	Actual	Obs. No.		
2.490	2.486	31	6.791	6.558	34	1.759	1.854	44	11.877	13.767	34		
2.353	2.336	7	5.025	4.785	44	1.757	1.770	48	5.406	5.402	44		
2.249	2.244	28	3.483	4.469	42	1.478	1.491	34	3.559	4.626	42		
2.028	2.021	5	1.405	1.260	48	1.199	1.144	43	2.993	3.650	43		
..	..	..	..	..	..	..	..	..	..	..	..		
0.248	0.143	44	0.001	0.001	53	−0.073	−0.085	49	0.016	0.017	24		
−0.108	−0.117	52	0.000	0.000	17	−0.090	−0.082	20	0.009	0.009	1		
−0.092	−0.113	37	0.000	0.000	24	0.039	0.045	37	0.000	0.001	30		
0.002	−0.014	11	0.000	0.000	21	−0.003	−0.010	10	0.000	0.000	53		

Table 10.3. Artistic qualities of painters: comparisons between estimated and actual influence of individual observations for the first two PCs, based on the correlation matrix.

Component number	1						2					
	Influence						Influence					
	Eigenvalue			Eigenvector			Eigenvalue			Eigenvector		
	Estimated	Actual	Obs. No.	Estimated	Actual	Obs. No.	Estimated	Actual	Obs. No.	Estimated	Actual	Obs. No.
	−0.075	−0.084	34	1.406	1.539	26	0.075	0.080	34	5.037	7.546	34
	0.068	0.074	31	1.255	1.268	44	0.064	0.070	44	4.742	6.477	43
	−0.067	−0.073	43	0.589	0.698	42	0.061	0.062	43	4.026	4.333	26
	0.062	0.067	28	0.404	0.427	22	0.047	0.049	39	1.975	2.395	42
	⋮	⋮	⋮	⋮	⋮	⋮	⋮	⋮	⋮	⋮	⋮	⋮
	−0.003	−0.003	6	0.001	0.001	33	−0.001	−0.002	45	0.011	0.012	21
	−0.001	−0.001	32	0.000	0.000	53	−0.001	−0.001	14	0.010	0.012	38
	0.001	0.001	11	0.000	0.000	30	−0.000	−0.001	42	0.001	0.001	30
	0.001	0.001	13	0.000	0.000	21	−0.000	−0.000	11	0.001	0.001	53

In considering changes to an eigenvector, there are changes to each of the p $(=4)$ coefficients in the vector. Comparing vectors is more difficult than comparing scalars, but Tables 10.2 and 10.3 give the sum of squares of changes in the individual coefficients of each vector, which is a plausible measure of the difference between two vectors. This quantity is a monotonic increasing function of the angle, in p-dimensional space, between the original and perturbed versions of $\mathbf{a}_k$, which further increases its plausibility. The idea of using angles between eigenvectors to compare PCs is discussed in a different context in Section 11.5.

The 'actual' changes for eigenvectors again come from leaving out one observation at a time, recomputing and then comparing the PCs, while the estimated changes are computed from multiples of sample versions of the expression (10.2.4) for $I(\mathbf{x}; \boldsymbol{\alpha}_k)$, and the corresponding expression for correlation matrices which is given by Calder (1985). The changes in eigenvectors derived in this way are much smaller in absolute terms than the changes in eigenvalues, so the eigenvector changes have been multiplied by 10^3 in Tables 10.2 and 10.3 in order that all the numbers are of comparable size.

A number of comments can be made regarding the results given in Tables 10.2 and 10.3. First, the estimated values are extremely good, in terms of obtaining the correct ordering of the observations with respect to their influence. There are some moderately large discrepancies in absolute terms for the observations with the largest influences, but experience with this and several other data sets suggests that the most influential observations are correctly identified, unless sample sizes become very small. The discrepancies in absolute values can also be reduced—by taking multiples other than $1/(n-1)$ and by including second-order (ε^2) terms.

A second point is that the observations which are most influential for a particular eigenvalue need not be so for the corresponding eigenvector, and vice versa. For example, there is no overlap between the four most influential observations for the first eigenvalue and its eigenvector, in either the correlation or covariance matrix. Conversely, observations can sometimes have a large influence on both an eigenvalue and its eigenvector—see Table 10.3, component 2, observation 34.

Next, note that observations may be influential for one PC only, or affect two or more. An observation is least likely to affect more than one PC in the case of eigenvalues for a covariance matrix—indeed there is no overlap between the four most influential observations for the first and second eigenvalues in Table 10.2. However, for eigenvalues in a correlation matrix, more than one value is likely to be affected by a very influential observation, because the sum of eigenvalues remains fixed. Also, large changes in an eigenvector for either correlation or covariance matrices will result in at least one other eigenvector being similarly changed, because of the orthogonality constraints. These results are again reflected in Tables 10.2 and 10.3, with observations appearing as influential for both of the first two eigenvectors, and for both eigenvalues in the case of the correlation matrix.

Comparing the results for covariance and correlation matrices in Tables 10.2 and 10.3, we see that several observations are influential for both matrices. This agreement has occurred because, in the present example, the original variables all have similar variances, so that the PCs for correlation and covariance matrices are very similar. In examples where the PCs based on correlation and covariance matrices are very different, the sets of influential observations for the two analyses often show little overlap.

Turning now to the observations which have been identified as influential in Table 10.3, we can examine their positions with respect to the first two PCs on Figures 5.2 and 5.3. Observation 34 which is the most influential observation on eigenvalues 1 and 2 and on eigenvector 2 is the painter indicated in the top left of Figure 5.2, Fr. Penni. His position is not particularly extreme with respect to the first PC, and he does not have an unduly large influence on its direction. However, he does have a strong influence on both the direction and variance (eigenvalue) of the second PC and to balance the increase which he causes in the second eigenvalue there is a compensatory decrease in the first eigenvalue. Hence, he is influential on that eigenvalue too. Observation 43, Rembrandt, is at the bottom of Figure 5.2, and, like Fr. Penni, has a direct influence on PC2 with an indirect, but substantial, influence on the first eigenvalue. The other two observations, 28 and 31, Caravaggio and Palma Vecchio, which are listed in Table 10.3 as being influential for the first eigenvalue, have a more direct effect. They are the two observations with the most extreme values on the first PC and appear at the extreme left of Figure 5.2.

Finally, the observations in Table 10.3 which are most influential on the first eigenvector, two of which also have large values of influence for the second eigenvector, appear on Figure 5.2 in the second and fourth quadrants, in moderately extreme positions.

Student Anatomical Measurements

In the discussion of the data on student anatomical measurements earlier in this chapter, it was suggested that observation 16 was so extreme on the second PC that it could be largely responsible for the direction of that component. Looking at influence functions for these data will enable us to consider this conjecture. Not surprisingly, observation 16 is the most influential observation for the second eigenvector, and, in fact, has an influence nearly six times as large as that of the second most influential observation— it is also the most influential on the first, third and fourth eigenvectors, showing that the perturbation caused by observation 16 to the second PC has a 'knock-on' effect to other PCs in order to preserve orthogonality. Although observation 16 is very influential on the eigenvectors, its effect is less marked on the eigenvalues. It has only the fifth highest influence on the second eigenvalue, though it is highest for the fourth eigenvalue, second

highest for the first, and fourth highest for the third. This behaviour is probably due to the constraint that the eigenvalues of a correlation matrix sum
to a fixed constant, and the consequent compensatory decreases in some
eigenvalues which follow on an increase in one eigenvalue. It is clear that
values of influence on eigenvalues may not fully reflect major changes in the
structure of the eigenvectors, at least when dealing with correlation matrices.

Having said that observation 16 is clearly the most influential, for eigenvectors, of the 28 observations in the data set, it should be noted that its
influence in absolute terms is not outstandingly large. In particular, the
coefficients, rounded to one decimal place, for the second PC, when observation 16 is omitted, are

$$0.2 \quad 0.1 \quad -0.4 \quad 0.8 \quad -0.1 \quad -0.3 \quad -0.2.$$

The corresponding coefficients when all 28 observations are included are

$$-0.0 \quad -0.2 \quad -0.2 \quad 0.9 \quad -0.1 \quad -0.0 \quad -0.0.$$

Thus, when observation 16 is removed, the basic character of PC2, as mainly
a measure of head size, is unchanged, although the dominance of head size in
this component is reduced. The angle between the two vectors defining the
second PCs, with and without observation 16, is about 24°, which is perhaps
larger than would be deduced from a quick inspection of the simplified
coefficients above.

10.3. Robust Estimation of Principal Components

One of the criticisms made in Section 3.7 of most of the available inference
procedures for PCs is that they depend on the often unrealistic assumption of
multivariate normality for x. If all that is required is a description of the
sample as given by its PCs, or a straightforward use of sample PCs as point
estimates of the population PCs, then the form of the distribution of x is
usually not very important. The main exception to this statement is when
outliers may occur. If the outliers are, in fact, influential observations, they
will have a disproportionate effect on the PCs, and if PCA is used blindly in
this case, without considering whether any observations are influential, then
the results can be largely determined by such observations. For example,
suppose that all but one of the n observations lie very close to a plane within
p-dimensional space, so that there are two dominant PCs for these $(n-1)$
observations. If the distance of the remaining observation from the plane is
greater than the variation of the $(n-1)$ observations within the plane, then
the first component for the n observations will be determined solely by the
direction of the single outlying observation from the plane. This, incidentally,
is a case where the first PC (rather than last few) will detect the outlier (see

Section 10.1), but if the distance from the plane is larger than distances within the plane, then the observation is likely to 'stick out' on at least one of the original variables as well.

To avoid such problems, it is possible to use robust estimation of the covariance or correlation matrix, and hence of the PCs. Such estimation downweights or discards the effect of any outlying observations, and five robust estimators have been investigated by Devlin *et al.* (1981). These five estimators use three different approaches, the first being to robustly estimate each element of the covariance or correlation matrix separately and then to 'shrink' the elements to achieve positive-definiteness, if this is not achieved with the initial estimates. The second type of approach involves robust regression of x_j on $(x_1, x_2, \ldots, x_{j-1})$ for $j = 2, 3, \ldots, p$. An illustration of the robust regression approach is presented, for a two-variable example, by Cleveland and Guarino (1976). Finally, the third approach has three variants; in one, the observations with the largest Mahalanobis distances from a robust estimate of the mean of **x** are discarded, and in the other two they are down-weighted. One of these variants is used by Coleman (1985) in the context of quality control.

All but the first of these five estimates involve iteration; full details are given by Devlin *et al.* (1981), who also show that the usual estimator of the covariance or correlation matrix can lead to misleading PCs if outlying observations are included in the analysis. Of the five possible robust alternatives which they investigated, only one, based on robust regression, is clearly dominated by other methods, and each of the remaining four may be chosen in some circumstances.

Robust estimation using an iterative procedure, based on downweighting observations with large Mahalanobis distance from the mean, was independently described by Campbell (1980). He also noted that, as a by-product of robust estimation, the weights given to each data point will give an indication of potential outliers. Since the weights are non-increasing functions of Mahalanobis distance, this procedure is essentially using the statistic d_{2i}^2, defined in Section 10.1, to identify outliers, except that the mean and covariance matrix of **x** are estimated robustly in the present case.

Campbell (1980) also proposed a modification of this robust estimation technique, in which the weights assigned to an observation, in estimating the covariance matrix, are no longer functions of the overall Mahalanobis distance of each observation from the mean. Instead, when estimating the kth PC, the weights are a decreasing function of the absolute value of the score of each observation for the kth PC. As with most of the techniques tested by Devlin *et al.* (1981), the procedure is iterative and the algorithm to implement it is quite complicated, with nine separate steps. However, Campbell (1980) and Matthews (1984) each give an example for which the technique works well, and as well as estimating the PCs robustly, both authors use the weights found by the technique to identify potential outliers.

The discussion in the previous section of this chapter noted that observations need not be particularly 'outlying' in order to be influential. Thus, robust estimation methods which give weights based on Mahalanobis distance to each observation, will 'miss' any influential observations which are not extreme with respect to Mahalanobis distance. It would seem preferable to downweight observations according to their influence, rather than their Mahalanobis distance. As yet, no systematic work seems to have been done on this idea, but it should be noted that the influence function for the kth eigenvalue of a covariance matrix is an increasing function of the absolute score on the kth PC. The weights used in Campbell's (1980) procedure are therefore downweighting observations according to their influence on eigenvalues (though not eigenvectors) of the covariance (but not correlation) matrix.

A different type of approach to the robust estimation of PCs is discussed by Gabriel and Odoroff (1983). The approach relies on the fact that PCs may be computed from the singular value decomposition (SVD) of the $(n \times p)$ data matrix X—see Section 3.5 and Appendix A1. To find the SVD, and hence the PCs, a set of equations, involving weighted means of functions of the elements of X, must be solved iteratively. Replacing the weighted means by medians, weighted trimmed means, or some other measure of location which is more robust than the mean, will lead to estimates of PCs which are less sensitive than the usual estimates to the presence of 'extreme' observations.

Yet another approach, based on 'projection pursuit', has been proposed by Li and Chen (1985). As with Gabriel and Odoroff (1983), and unlike Campbell (1980) and Devlin et al. (1981), the PCs are estimated directly without first finding a robustly estimated covariance matrix. Indeed, Li and Chen (1985) suggest that it may be better to estimate the covariance matrix, Σ, from the robust PCs, via the spectral decomposition (see Property A3 of Sections 2.1 and 3.1), rather than estimating Σ directly. Li and Chen's (1985) idea is to find linear functions of x which maximize a robust estimate of scale, rather than maximize variance. Properties of their estimates are investigated, and their empirical behaviour is compared with that of Devlin et al. (1981)'s estimates, using simulation studies. Similar levels of performance are found for both types of estimate.

A different, but related, topic is robust estimation of the *distribution* of the PCs and their variances, rather than robust estimation of the PCs themselves. It was noted in Section 3.6 that this can be done using bootstrap estimation (Diaconis and Efron, 1983). The 'shape' of the estimated distributions should also give some indication of whether any highly influential observations are present in a data set (the distributions may be multimodal, corresponding to the presence or absence of the influential observations in each sample from the data set), although the method will not directly identify such observations.

Finally, we mention that Ruymgaart (1981) discusses a class of robust PC estimators. However, his discussion is restricted to bivariate distributions (i.e. $p = 2$) and is entirely theoretical in nature.

10.4. Concluding Remarks

The topics discussed in this chapter pose difficult problems in data analysis, and research is continuing on all of them. It is, of course, useful to identify potentially outlying observations and PCA provides a number of ways of doing so. Similarly, it is important to know which observations have the greatest influence on the results of a PCA.

Identifying potential outliers and influential observations is, however, only part of the problem; the next, perhaps more difficult task, is to decide whether the most extreme or influential observations are, in fact, sufficiently extreme or influential to warrant further action, and, if so, what that action should be. Tests of significance for outliers were discussed only briefly in Section 10.1, because they are usually only approximate, and tests of significance for influential observations in PCA are at an early stage of development. Perhaps the best advice is that observations which are much more extreme or influential than all (or most, to avoid masking by other extreme observations) of the remaining observations in the data set, should be thoroughly investigated, and explanations sought for their behaviour. The analysis could also be repeated with such observations omitted, although it may be dangerous to act as if the deleted observations never existed. Robust estimation provides an automatic way of dealing with extreme (or influential) observations, but, if at all possible, it should be accompanied by an examination of any observations which have been omitted or substantially downweighted by the analysis.

CHAPTER 11

Principal Component Analysis for Special Types of Data

In much of statistical inference, it is assumed that a data set consists of n *independent* observations on one or more random variables, $\mathbf{x}$, and this assumption is often implicit when a PCA is done. Another assumption which also may be made implicitly is that $\mathbf{x}$ consists of *continuous* variables, with perhaps the stronger assumption of multivariate normality if we require to make some formal inference for the PCs.

If PCA is to be used only as a descriptive, rather than inferential, tool, then none of these assumptions is essential in order for the technique to be valuable. However, relaxation of one or more of the assumptions may imply that 'ordinary' PCA can be usefully adapted or extended in order to gain a greater appreciation of the structure of the data. The present chapter discusses the implications for PCA of several special types of data in which standard assumptions do not hold.

Section 11.1 looks at a number of ideas involving PCA for discrete data. In particular, correspondence analysis, which has already been mentioned as a graphical technique in Section 5.4, is discussed further.

One of the most important situations in which independence does not hold is when the data consist of p time series (each of length n); such data, and several relevant modifications of PCA, are described in Section 11.2. Some of the modifications involve quite complicated mathematics and only the briefest of outlines is given.

In Section 11.3, we discuss compositional data, in which the p elements of $\mathbf{x}$ are constrained to sum to the same constant (usually 1 or 100) for all observations, and, in Section 11.4, some uses of PCA for data from designed experiments are described.

In Section 11.5, we look at the possibility of defining 'common' PCs when the observations come from several distinct populations, and also examine ways of comparing PCs from the different populations.

Section 11.6 discusses possible ways of dealing with missing data in a PCA, and the final section (Section 11.7) describes how (somewhat nonstandard) PCs can be used in testing the goodness-of-fit of a probability distribution to a set of data.

11.1. Principal Component Analysis for Discrete Data

When PCA is used as a descriptive technique, there is no reason for the variables in the analysis to be of any particular type. At one extreme, $\mathbf{x}$ may have a multivariate normal distribution, in which case all the relevant inferential results mentioned in Section 3.7 may be used. At the opposite extreme, the variables could be a mixture of continuous, ordinal or even binary (0–1) variables. It is true that variances, covariances and correlations have especial relevance for multivariate normal $\mathbf{x}$, and that linear functions of binary variables are less readily interpretable than linear functions of continuous variables. However, the basic objective of PCA, to summarize most of the 'variation' which is present in the original set of p variables, using a smaller number of derived variables, can be achieved regardless of the nature of the original variables.

For data in which all variables are binary, Gower (1966) points out that using PCA *does* provide a plausible low-dimensional representation. This follows because PCA is equivalent to a principal co-ordinate analysis based on the commonly used definition of similarity between two individuals (observations) as the proportion of the p variables for which the two individuals take the same value (see Section 5.2). Cox (1972), however, suggests an alternative to PCA for binary data. His idea, which he calls 'permutational principal components', is based on the fact that a set of data consisting of p binary variables can be expressed in a number of different but equivalent ways. As an example, consider the following two variables from a cloud-seeding experiment:

$$x_1 = \begin{cases} 1 & \text{if rain falls in seeded area,} \\ 0 & \text{if no rain falls in seeded area,} \end{cases}$$

$$x_2 = \begin{cases} 1 & \text{if rain falls in control area,} \\ 0 & \text{if no rain falls in control area.} \end{cases}$$

Instead of x_1, x_2 we could define

$$x_1' = \begin{cases} 1 & \text{if both areas have rain or both areas dry,} \\ 0 & \text{if one area has rain, the other is dry,} \end{cases}$$

and $x_2' = x_1$. There is also a third possibility namely $x_1'' = x_1'$, $x_2'' = x_2$. For $p > 2$ variables there are many more possible permutations of this type, and Cox (1972) suggests that an alternative to PCA might be to transform to *independent* binary variables using such permutations. Bloomfield (1974) investigates Cox's suggestion in more detail, and presents an example which has four variables. He notes that in examples involving two variables we can write

$$x'_1 = x_1 + x_2 \text{ (modulo 2)}, \quad \text{using the notation above.}$$

For more than two variables, not all permutations can be rewritten in this way, but Bloomfield restricts his attention to those permutations which can. Thus, for a set of p binary variables $x_1, x_2, \ldots, x_p$, he considers transformation to $z_1, z_2, \ldots, z_p$ such that, for $k = 1, 2, \ldots, p$, we have either

$$z_k = x_j \quad \text{for some } j,$$

or

$$z_k = x_i + x_j \text{ (modulo 2)} \quad \text{for some } i, j, \quad i \neq j.$$

He is thus restricting attention to *linear* transformations of the variables (as in PCA) and the objective in this case is to choose a transformation which simplifies the structure between variables. The data can be viewed as a contingency table, and Bloomfield (1974) interprets a simpler structure as one which reduces high-order interaction terms between variables. This idea is illustrated on a four-variable example, and several transformations are examined, but (unlike PCA) there is no algorithm for finding a unique 'best' transformation.

A second special type of discrete data occurs when, for each observation, only ranks, and not actual values, are given for each variable. For such data, Gower (1967) points out that all the rows of the data matrix, $\mathbf{X}$, have the same sum, so that the data are constrained in a way which is similar to that which holds for compositional data (see Section 11.3). Gower (1967) discusses some geometric implications which follow from the constraints imposed by this type of ranked data and by compositional data.

Another possible adaptation of PCA to discrete data might be to replace variances and covariances by measures of dispersion and association which are more relevant to discrete variables. For the particular case of contingency table data, many different measures of association have been suggested (Bishop *et al.*, 1975, Chapter 11). It is also possible to define measures of variation other than variance for such data, for example, Gini's measure (see Bishop *et al.*, 1975, Section 11.3.4). It should be possible to adapt PCA to optimize such alternative measures, but the more usual 'adaptation' of PCA for contingency table data is correspondence analysis. This technique, together with an alternative 'non-metric' PCA described by De Leeuw and van Rijckevorsel (1980), will be the subject of the remainder of this section.

Correspondence analysis has already been discussed briefly in Section 5.4 as a graphical technique. It is appropriate to discuss correspondence analysis in a little more detail in the present section because, as well as being used as a graphical means of displaying contingency table data (see Section 5.4), the technique has been described by some authors as a form of PCA for nominal data, e.g. De Leeuw and van Rijckevorsel (1980). To see in what way this description is valid, consider, as in Section 5.4, a data set of $n_{..}$ observations arranged in a two-way contingency table, with n_{ij} denoting the number of observations which take the ith value for the first (row) variable and the jth value for the second (column) variable, $i = 1, 2, \ldots, r; j = 1, 2, \ldots, c$. Let $\mathbf{N}$ be the $(r \times c)$ matrix with (i, j)th element n_{ij}, and define $\mathbf{P} = (1/n_{..})\mathbf{N}$, $\mathbf{r} = \mathbf{Pl}_c$, $\mathbf{c} = \mathbf{P'l}_r$, and $\mathbf{X} = \mathbf{P} - \mathbf{rc'}$, where $\mathbf{l}_c$, $\mathbf{l}_r$ are vectors of c and r elements respectively, with all elements unity. If the variable defining the rows of the contingency table is independent of the variable defining the columns, then the matrix of 'expected counts' will be given by $n_{..}\,\mathbf{rc'}$. Thus, $\mathbf{X}$ is a matrix of the residuals which occur when the 'independence' model is fitted to $\mathbf{P}$.

The generalized singular value decomposition (SVD) of $\mathbf{X}$, is defined by

$$\mathbf{X} = \mathbf{VMB'}, \tag{11.1.1}$$

where $\mathbf{V'\Omega V} = \mathbf{I}$, $\mathbf{B'\Phi B} = \mathbf{I}$, $\mathbf{M}$ is diagonal, and $\mathbf{I}$ is the identity matrix whose dimension equals the rank of $\mathbf{X}$ (see also Section 12.1). If $\mathbf{\Omega} = \mathbf{D}_r^{-1}$, $\mathbf{\Phi} = \mathbf{D}_c^{-1}$, where $\mathbf{D}_r$, $\mathbf{D}_c$ are diagonal matrices whose diagonal entries are the elements of $\mathbf{r}$, $\mathbf{c}$ respectively, then the columns of $\mathbf{B}$ define principal axes for the set of r 'observations' given by the rows of $\mathbf{X}$. Similarly, the columns of $\mathbf{V}$ define principal axes for the set of c 'observations' given by the columns of $\mathbf{X}$, and from the first q columns of $\mathbf{B}$ and $\mathbf{V}$, respectively, we can derive the 'co-ordinates' of the rows and columns of $\mathbf{N}$ in q-dimensional space (see Greenacre, 1984, p. 87) which are the end-products of a correspondence analysis.

A correspondence analysis is therefore based on a generalized SVD of $\mathbf{X}$ and, as will be shown in Section 12.1, this is equivalent to an 'ordinary' SVD of

$$\tilde{\mathbf{X}} = \mathbf{\Omega}^{1/2}\mathbf{X}\mathbf{\Phi}^{1/2}$$

$$= \mathbf{D}_r^{-1/2}\mathbf{X}\mathbf{D}_c^{-1/2}$$

$$= \mathbf{D}_r^{-1/2}\left(\frac{1}{n_{..}}\mathbf{N} - \mathbf{rc'}\right)\mathbf{D}_c^{-1/2}.$$

The SVD of $\tilde{\mathbf{X}}$ can be written

$$\tilde{\mathbf{X}} = \mathbf{WKC'}, \tag{11.1.2}$$

with $\mathbf{V}$, $\mathbf{M}$, $\mathbf{B}$ of (11.1.1) defined in terms of $\mathbf{W}$, $\mathbf{K}$, $\mathbf{C'}$ of (11.1.2) as

$$\mathbf{V} = \mathbf{\Omega}^{-1/2}\mathbf{W}, \qquad \mathbf{M} = \mathbf{K}, \qquad \mathbf{B} = \mathbf{\Phi}^{-1/2}\mathbf{C}.$$

If we consider $\tilde{\mathbf{X}}$ as a matrix of r observations on c variables, then the coefficients of the PCs for $\tilde{\mathbf{X}}$ are given in the columns of $\mathbf{C}$, and the co-ordinates (scores) of the observations with respect to the PCs are given by the elements of $\mathbf{WK}$ (see the discussion of the biplot with $\alpha = 1$ in Section 5.3).

Thus, the positions of the row points given by correspondence analysis are rescaled versions of the values of the PCs for the matrix $\tilde{\mathbf{X}}$. Similarly, the column positions given by correspondence analysis are rescaled versions of values of PCs for the matrix $\tilde{\mathbf{X}}'$, a matrix of c observations on r variables. In this sense, correspondence analysis can be thought of as a form of PCA for a transformation $\tilde{\mathbf{X}}$ of the original contingency table $\mathbf{N}$ (or a generalized PCA for $\mathbf{X}$—see Section 12.1).

Because of the various optimality properties of PCs discussed in Chapters 2 and 3, and also the fact that the SVD provides a sequence of 'best-fitting' approximations to $\tilde{\mathbf{X}}$ of rank 1, 2, ..., as defined below equation (3.5.3), it follows that correspondence analysis provides co-ordinates for rows and columns of $\mathbf{N}$ which give the best fit, in a small number of dimensions (usually two), to a transformed version, $\tilde{\mathbf{X}}$, of $\mathbf{N}$. This rather convoluted definition of correspondence analysis has been used because of its connection with PCA, but there are a number of other definitions, which turn out to be equivalent, as shown in Greenacre (1984, Chapter 4). In particular, the techniques of reciprocal averaging and dual (or optimal) scaling are widely used in ecology and psychology, respectively. The rationale behind each technique is different, and differs in turn from that of correspondence analysis, but numerically all three techniques provide the same results for a given table of data.

Returning to the definition of correspondence analysis discussed above, the centring of $\mathbf{P}$ by the subtraction of $\mathbf{rc}'$ is, in fact, unnecessary. If the generalized SVD is found directly for $\mathbf{P}$, rather than for $\mathbf{X} = \mathbf{P} - \mathbf{rc}'$, then the only change is to add an extra, trivial dimension, with the other dimensions remaining unchanged. Thus, correspondence analysis can also be thought of as a generalized PCA for $\mathbf{P}$, or a PCA for $\mathbf{D}_r^{-1/2}\mathbf{PD}_c^{-1/2}$, in both cases ignoring the first (trivial) PC. In this sense, then, correspondence analysis is a form of PCA for two-way contingency table data, i.e. for nominal data with two variables.

The ideas of correspondence analysis can be extended to contingency tables involving more than two variables (Greenacre, 1984, Chapter 5), and links with PCA remain in this case, as will now be discussed very briefly. Instead of doing a correspondence analysis using the $(r \times c)$ matrix $\mathbf{N}$, it is possible to use a $[n_{..} \times (r + c)]$ indicator matrix $\mathbf{Z} = (\mathbf{Z}_1 \quad \mathbf{Z}_2)$ where $\mathbf{Z}_1$ is $(n_{..} \times r)$ and has (i, j)th element equal to 1 if the ith observation takes the jth value for the first (row) variable, and zero otherwise. Similarly, $\mathbf{Z}_2$ is $(n_{..} \times c)$, with (i, j)th element equal to 1 if the ith observation takes the jth value for the second (column) variable and zero otherwise. If we have a contingency table with more than two variables, then we can extend the analysis based on $\mathbf{Z}$, leading to 'multiple correspondence analysis' (see also Section 12.2), by adding further indicator matrices $\mathbf{Z}_3, \mathbf{Z}_4, ...,$ to $\mathbf{Z}$, one matrix for each additional variable. Another alternative to carrying out the analysis on $\mathbf{Z} = (\mathbf{Z}_1 \quad \mathbf{Z}_2 \quad \mathbf{Z}_3 ...)$, is to base the analysis on the so-called Burt matrix, $\mathbf{Z}'\mathbf{Z}$, instead (Greenacre, 1984, p. 140).

In the case where each variable can take only two values, Greenacre (1984, p. 145) notes two relationships between (multiple) correspondence analysis and PCA. He states that correspondence analysis of $\mathbf{Z}$ is closely related to PCA of a matrix $\mathbf{Y}$, whose ith column is one of the two columns of $\mathbf{Z}_i$, standardized to have unit variance. Furthermore, the correspondence analysis of the Burt matrix $\mathbf{Z'Z}$ is equivalent to a PCA of the correlation matrix $(1/n_{..})\mathbf{Y'Y}$. Thus, the idea of correspondence analysis as a form of PCA for nominal data is valid for any number of binary variables. A final relationship between correspondence analysis and PCA (Greenacre, 1984, p. 183) occurs when correspondence analysis is done for a special type of 'doubled' data matrix, when each variable is repeated twice, once in its original form, and the second time in a complementary form (for details, see Greenacre (1984, Chapter 6).)

We turn now to the 'non-metric' generalization of PCA for discrete data, which is a form of non-linear PCA (see Section 12.2). The method is based on a generalization of the result that if, for a data matrix $\mathbf{X}$, we minimize

$$\text{tr}\{(\mathbf{X} - \mathbf{YB'})'(\mathbf{X} - \mathbf{YB'})\}, \tag{11.1.3}$$

with respect to the $(n \times q)$ matrix $\mathbf{Y}$ whose columns are linear functions of columns of $\mathbf{X}$, and with respect to the $(q \times p)$ matrix $\mathbf{B'}$, where the columns of $\mathbf{B}$ are orthogonal, then the optimal $\mathbf{Y}$ consists of the values of the first q PCs for the n observations, and the optimal matrix $\mathbf{B}$ consists of the coefficients of the first q PCs. (In order to be consistent with Greenacre (1984), the symbols $\mathbf{X}, \mathbf{Y}, \mathbf{B}$ were used for different quantities earlier in this section. Here we return to their usage elsewhere in the present text.) The criterion (11.1.3) corresponds to that used in the sample version of Property A5 (see Section 2.1), and can be rewritten as

$$\text{tr}\left\{ \sum_{j=1}^{p} (\mathbf{x}_j - \mathbf{Yb}_j)'(\mathbf{x}_j - \mathbf{Yb}_j) \right\}, \tag{11.1.4}$$

where $\mathbf{x}_j, \mathbf{b}_j$ are the jth columns of $\mathbf{X}, \mathbf{B'}$ respectively; minimizing (11.1.4) is, in turn, equivalent to minimizing

$$\sum_{i=1}^{p} \text{tr}(\mathbf{Y} - \mathbf{x}_j\mathbf{b}_j')'(\mathbf{Y} - \mathbf{x}_j\mathbf{b}_j') \tag{11.1.5}$$

(De Leeuw and van Rijckevorsel, 1980). Generalizations of the criterion (11.1.5) can be defined by introducing diagonal matrices $\mathbf{M}_j$ to give

$$\sum_{j=1}^{p} \text{tr}\{(\mathbf{Y} - \mathbf{x}_j\mathbf{b}_j')'\mathbf{M}_j(\mathbf{Y} - \mathbf{x}_j\mathbf{b}_j')\} \tag{11.1.6}$$

and by relaxing the requirement that the columns of $\mathbf{Y}$ are *linear* functions of the columns of $\mathbf{X}$. De Leeuw and van Rijckevorsel (1980) indicate that different choices of $\mathbf{M}_j$ in (11.1.6), and different choices of spaces for $\mathbf{Y}$, are appropriate for different types of discrete data. These choices have been implemented in a computer program PRINCALS (Gifi, 1983), but published examples on real data sets are not easy to find.

11.2. Principal Component Analysis for Non-independent and Time Series Data

In much of statistics it is assumed that the n observations $x_1, x_2, \ldots, x_n$ are independent. This section discusses first the general implications for PCA of non-independence among $x_1, x_2, \ldots, x_n$, but most of the section is concerned with PCA for time series data, the most common type of non-independent data. Time series data are sufficiently different from ordinary independent data for there to be several aspects of PCA which arise only for such data, for example, PCs in the frequency domain, and 'complex' PCs; these topics will be included in the present section.

The results of Section 3.7 which allow formal inference procedures to be performed for PCs rely on independence of $x_1, x_2, \ldots, x_n$, as well as on multivariate normality. They cannot therefore be used if more than very weak dependence is present between $x_1, x_2, \ldots, x_n$. However, as stated several times before, the main objective of PCA is often descriptive, not inferential, and complications such as non-independence do not seriously affect this objective. The effective sample size is reduced below n, but this reduction need not be too important. In fact, in some circumstances we are actively looking for dependence among $x_1, x_2, \ldots, x_n$. For example, grouping of observations in a few small areas of the two-dimensional space defined by the first two PCs implies dependence between those observations which are grouped together. Such behaviour is actively sought in cluster analysis (see Section 9.2), and would often be welcomed as a useful insight into the structure of the data, rather than decried as an undesirable feature. ·

In time series data, dependence between the x's is induced by their relative closeness in time, so that x_h and x_i will often be highly dependent if $|h - i|$ is small, with decreasing dependence as $|h - i|$ increases. This basic pattern may, in addition, be perturbed by, for example, seasonal dependence in monthly data, where decreasing dependence for increasing $|h - i|$ is interrupted by a higher degree of association for observations separated by exactly 1 year, 2 years, and so on.

We now need to introduce some of the basic ideas of time series analysis, although limited space permits only a rudimentary introduction to this vast subject (for more information see, for example, Brillinger (1981), Chatfield (1984) or Priestley (1981)). Suppose, for the moment, that only a single variable is measured, at equally-spaced points in time. Our time series is then $\ldots x_{-1}, x_0, x_1, x_2, \ldots$. Much of time series analysis is concerned with series which are stationary, and which can be described entirely by their first- and second-order moments; these moments are

$$\mu = E(x_i), \qquad\qquad i = \ldots, -1, 0, 1, 2, \ldots,$$

$$\gamma_k = E[(x_i - \mu)(x_{i+k} - \mu)], \qquad i = \ldots, -1, 0, 1, 2, \ldots; \qquad (11.2.1)$$

$$k = \ldots, -1, 0, 1, 2, \ldots,$$

μ is the mean of the series, and is the same for all x_i in stationary series, and γ_k, the kth autocovariance, is the covariance between x_i and x_{i+k}, which depends on k, but not i, for stationary series. The information contained in the autocovariances can be expressed equivalently in terms of the spectrum of the series

$$f(\lambda) = \frac{1}{2\pi} \sum_{k=-\infty}^{\infty} \gamma_k e^{-ik\lambda}, \tag{11.2.2}$$

where $i = \sqrt{-1}$. Roughly speaking, the function $f(\lambda)$ decomposes the series into oscillatory portions with different frequencies of oscillation, and $f(\lambda)$ measures the relative importance of these portions as a function of their angular frequency λ. For example, if a series is almost a pure oscillation with angular frequency λ_0, then $f(\lambda)$ will be large for λ close to λ_0 and near zero elsewhere. This behaviour would be signalled in the autocovariances by a large value of γ_k at $k = k_0$, where k_0 is the period of oscillation corresponding to angular frequency λ_0 (i.e. $k_0 = 2\pi/\lambda_0$), and small values elsewhere.

Because there are two different, but equivalent, functions (11.2.1) and (11.2.2) which express the second-order behaviour of a time series, there are two different types of analysis of time series, namely in the time domain using (11.2.1), or in the frequency domain using (11.2.2).

Consider now a time series which consists not of a single variable, but p variables. The definitions (11.2.1), (11.2.2) generalize readily to

$$\mathbf{\Gamma}_k = E[(\mathbf{x}_i - \mathbf{\mu})(\mathbf{x}_{i+k} - \mathbf{\mu})'], \tag{11.2.3}$$

where

$$\mathbf{\mu} = E[\mathbf{x}_i],$$

and

$$\mathbf{F}(\lambda) = \frac{1}{2\pi} \sum_{k=-\infty}^{\infty} \mathbf{\Gamma}_k e^{-ik\lambda}, \tag{11.2.4}$$

$\mathbf{\Gamma}_k$ and $\mathbf{F}(\lambda)$ are now $(p \times p)$ matrices.

Principal component analysis operates on a covariance or correlation matrix, but in time series we can not only calculate covariances between variables measured at the same time (the usual definition of covariance, which is given by the matrix $\mathbf{\Gamma}_0$ defined in (11.2.3)), but also covariances between variables at different times, as measured by $\mathbf{\Gamma}_k$, $k \neq 0$. This is in contrast to the more usual situation where our observations $\mathbf{x}_1, \mathbf{x}_2, \ldots$ are independent, so that any covariances between elements of $\mathbf{x}_i$, $\mathbf{x}_j$ are zero when $i \neq j$. In addition to the choice of which $\mathbf{\Gamma}_k$'s to examine, the fact that the covariances have an alternative representation in the frequency domain means that there are several different ways in which PCA can be applied to time series data. We now examine a number of these possibilities.

First, consider the straightforward use of PCA on a set of time series data $\mathbf{x}_1, \mathbf{x}_2, \ldots, \mathbf{x}_n$ as if it were an 'ordinary' data set of n observations on p

variables. We have already seen several examples of this kind of data in previous chapters. Section 4.3 gave an example of a type which is common in meteorology, where the variables are p different meteorological measurements made at the same geographical location, or measurements of the same meteorological variable made at p different geographical locations, and the n observations on each variable correspond to different times. The examples given in Section 4.5 and Section 6.4.2 are also illustrations of PCA applied to data for which the variables (stock prices and crime rates respectively) are measured at various points of time. Furthermore, one of the earliest published applications of PCA (Stone, 1947) was on (economic) time series data.

Next, suppose that our p-element random vector $\mathbf{x}$ consists of p consecutive values of a single time series, i.e. the time-variation is represented not by different observations on $\mathbf{x}$, but by different elements within $\mathbf{x}$. If the time series is stationary, the covariance matrix, Σ, for $\mathbf{x}$ will then have a special form in which the (i, j)th element depends only on $|i - j|$. Such matrices, known as Toeplitz matrices, have a well-known structure for their eigenvalues and eigenvectors (Brillinger, 1981, p. 108). Craddock (1965) performed an analysis of this type on monthly mean temperatures for central England for the period November 1680 to October 1963. The $p\ (= 12)$ elements of $\mathbf{x}$ are mean temperatures for the 12 months of a particular year, where a 'year' starts in November. There is some dependence between different values of $\mathbf{x}$, but it is weaker than that between elements within a particular $\mathbf{x}$; between-year correlation was minimized by starting each year at November, when there was apparently evidence of very little continuity in atmospheric behaviour. The covariance matrix does not, of course, have exact Toeplitz structure, but several of the eigenvectors have approximately the form expected for such matrices.

Durbin (1984) uses the structure of the eigenvectors of Toeplitz matrices in a different context. In regression analysis (see Chapter 8), if the dependent variable and the predictor variables are all time series, then Toeplitz structure for the covariance matrix of error terms in the regression model can be used to deduce properties of the least squares estimators of regression coefficients.

Reverting to 'standard' time series data, where $\mathbf{x}$ consists of p variables measured at the same time, PCA can be carried out in the frequency domain, and this clearly has no counterpart for data sets consisting of independent observations. Brillinger (1981, Chapter 9) devotes a whole chapter to PCA in the frequency domain—see also Priestley et al. (1974). To see how frequency domain PCs are derived, note that PCs for a p-variate random vector $\mathbf{x}$, with zero mean, can be obtained by finding $(p \times q)$ matrices $\mathbf{B}$, $\mathbf{C}$ such that

$$E[(\mathbf{x} - \mathbf{Cz})'(\mathbf{x} - \mathbf{Cz})]$$

is minimized, where $\mathbf{z} = \mathbf{B}'\mathbf{x}$. (This is equivalent to the criterion which defines Property A5 in Section 2.1.)

It turns out that $\mathbf{B} = \mathbf{C}$ and that the columns of $\mathbf{B}$ are the first q eigenvectors of $\mathbf{\Sigma}$, the covariance matrix of $\mathbf{x}$, so that the elements of $\mathbf{z}$ are the first q PCs for $\mathbf{x}$. This argument can be extended to a time series of p variables as follows. Suppose that our series is $\ldots \mathbf{x}_{-1}, \mathbf{x}_0, \mathbf{x}_1, \mathbf{x}_2, \ldots$ and that $E[\mathbf{x}_t] = \mathbf{0}$ for all t. Define

$$\mathbf{z}_t = \sum_{u=-\infty}^{\infty} \mathbf{B}'_{t-u} \mathbf{x}_u,$$

and estimate $\mathbf{x}_t$ by $\sum_{u=-\infty}^{\infty} \mathbf{C}_{t-u} \mathbf{z}_u$, where $\ldots \mathbf{B}_{t-1}, \mathbf{B}_t, \mathbf{B}_{t+1}, \mathbf{B}_{t+2}, \ldots, \mathbf{C}_{t-1}, \mathbf{C}_t,$ $\mathbf{C}_{t+1}, \mathbf{C}_{t+2}, \ldots$ are $(p \times q)$ matrices, which minimize

$$E\left[\left(\mathbf{x}_t - \sum_{u=-\infty}^{\infty} \mathbf{C}_{t-u} \mathbf{z}_u\right)^* \left(\mathbf{x}_t - \sum_{u=-\infty}^{\infty} \mathbf{C}_{t-u} \mathbf{z}_u\right)\right],$$

where $*$ denotes conjugate transpose. The difference between this formulation and that for ordinary PCs above, is that the relationships between $\mathbf{z}$ and $\mathbf{x}$ are in terms of all values of $\mathbf{x}_t$ and $\mathbf{z}_t$ at different times, rather than between a single $\mathbf{x}$ and $\mathbf{z}$. Also, the derivation is in terms of general complex, rather than restricted to real, series. It turns out (Brillinger, 1981, p. 344) that

$$\mathbf{B}'_u = \frac{1}{2\pi} \int_0^{2\pi} \tilde{\mathbf{B}}(\lambda) \, e^{iu\lambda} \, d\lambda,$$

$$\mathbf{C}_u = \frac{1}{2\pi} \int_0^{2\pi} \tilde{\mathbf{C}}(\lambda) \, e^{iu\lambda} \, d\lambda,$$

where $\tilde{\mathbf{C}}(\lambda)$ is a $(p \times q)$ matrix whose columns are the first q eigenvectors of the matrix $\mathbf{F}(\lambda)$ given in (11.2.4), and $\tilde{\mathbf{B}}(\lambda)$ is the conjugate transpose of $\tilde{\mathbf{C}}(\lambda)$.

The q series which form the elements of $\mathbf{z}_t$ are called the first q PC series of $\mathbf{x}_t$. Brillinger (1981, Sections 9.3, 9.4) discusses various properties and estimates of these PC series, and gives an example in Section 9.6 on monthly temperature measurements at 14 meteorological stations. Principal component analysis in the frequency domain has also been used on economic time series, for example on Dutch provincial unemployment data (Bartels, 1977, Section 7.7).

There is, in fact, a connection between frequency-domain PCs and PCs derived in the more usual way (Brillinger, 1981, Section 9.5). It is not a particularly straightforward connection, involving Hilbert transforms (see Brillinger, (1981, p. 32), for a definition) and the decomposition of $\mathbf{x}_t$ into elements corresponding to varying angular frequencies, λ, so details will not be given here.

Hilbert transforms also play a part in a technique which has appeared recently in the meteorological literature (Anderson and Rosen, 1983; Barnett, 1983). Suppose that the set of p series $\{\mathbf{x}_t\}$ consists of measurements of a meteorological variable at p different geographical locations. Each element of $\mathbf{x}_t$ is real but, in the new technique, it is augmented by an imaginary part which is proportional to the Hilbert transform of the element. The covariance matrix of the complex $2p$-element augmented vector is then computed, and PCs found for this matrix. These 'complex PCs' allow the variation in x_t to be decomposed in terms of *propagating* features in the atmosphere, rather than simply looking at *fixed* patterns such as are illustrated in Section 4.3.

Besse and Ramsay (1984) consider a set of time series data $\{x_t\}$ as a set of values of p smoothly varying functions. It may then be required to carry out PCA, but *taking into account* the smoothness of the functions which underlie the data. Besse and Ramsay (1984) use some quite sophisticated mathematics to show how this may be done. Rao (1964) prefers to *eliminate* any smooth trend terms which might dominate a PCA—see Section 12.5.

Yet another rôle for PCs in the analysis of time series data is presented by Doran (1976). In his paper, PCs are used to estimate the coefficients in a regression analysis of one time series variable on several others. The idea is similar to that of PC regression (see Section 8.1), but is more complicated since it involves the spectrum.

In this section a number of ways of using PCA (or adaptations of PCA) for time series data have been described briefly. Several of the techniques involve sophisticated mathematics, and have not yet been used extensively in practice. It is therefore difficult to evaluate the general usefulness of the techniques, although the originators of each technique have, in each case, given examples which illustrate the technique's potential value.

11.3. Principal Component Analysis for Compositional Data

Compositional data consists of observations $x_1, x_2, \ldots, x_n$, for which each element of x_i is a proportion, and the elements of x_i are constrained to sum to unity. Such data will occur, for example, when a number of chemical compounds, or geological specimens, or blood samples, are analysed, and the proportion, in each, of a number of chemical elements is recorded. As noted in Section 11.1, Gower (1967) discussed some geometric implications which follow from the constraints on the elements of x, but the major reference for PCA on compositional data is Aitchison (1983). Because of the constraints on the elements of x, and also because compositional data apparently often exhibit non-linear, rather than linear, structure among their variables, Aitchison (1983) has proposed that PCA should be modified for such data.

At first sight, it might seem that no 'real difficulty' is implied by the condition,

$$x_{i1} + x_{i2} + \cdots + x_{ip} = 1, \tag{11.3.1}$$

which holds for each observation. If a PCA is done on x, there will be a PC with zero eigenvalue which identifies the constraint. This PC can be ignored, because it is entirely predictable from the form of the data, and the remaining PCs can be interpreted as usual. A counter to this argument is that correlations and covariances, and hence PCs, cannot be interpreted in the usual way when the constraints is present. In particular, the constraint (11.3.1) introduces a bias towards negative values among the correlations, so that a set of compositional variables which are 'as independent as possible' will not all have zero correlations between them.

One way of overcoming this problem is to do the PCA on a subset of $(p - 1)$ of the p compositional variables, but this idea has the unsatisfactory feature that the choice of which variable to leave out is arbitrary, and different choices will lead to different PCs. For example, suppose that two variables have much larger variances than the other $(p - 2)$ variables. If a PCA is based on the covariance matrix for $(p - 1)$ of the variables, then the result will vary considerably, depending on whether the omitted variable has a large or small variance. Furthermore, there remains the restriction that the $(p - 1)$ chosen variables must sum to no more than unity, so that the interpretation of correlations and covariances is still not straightforward.

The alternative which is suggested by Aitchison (1983) is to replace $\mathbf{x}$ by $\mathbf{v} = \log[\mathbf{x}/g(\mathbf{x})]$, where $g(\mathbf{x}) = (x_1 x_2 \ldots x_p)^{1/p}$ is the geometric mean of the elements of $\mathbf{x}$. Thus, the jth element of $\mathbf{v}$ is

$$v_j = \log x_j - \frac{1}{p} \sum_{i=1}^{p} \log x_i, \qquad j = 1, 2, \ldots, p. \tag{11.3.2}$$

A PCA is then done for $\mathbf{v}$ rather than $\mathbf{x}$. There is one zero eigenvalue, having an eigenvector whose elements are all equal; the remaining eigenvalues are positive, and, because the corresponding eigenvectors are orthogonal to the final eigenvector, they define *contrasts* (i.e. linear functions whose coefficients sum to zero) for the log x_j's.

Aitchison (1983) also shows that these same functions can equivalently be found by basing a PCA on the non-symmetric set of variables $\mathbf{v}^{(j)}$, where

$$\mathbf{v}^{(j)} = \log[\mathbf{x}^{(j)}/x_j] \tag{11.3.3}$$

and $\mathbf{x}^{(j)}$ is the $(p - 1)$ vector obtained by deleting the jth element x_j from $\mathbf{x}$. The idea of transforming to logarithms before doing the PCA can, of course, be used for data other than compositional data—see also Section 12.2. However, there are a number of particular advantages of the log-ratio transformation (11.3.2), or equivalently (11.3.3), for compositional data. These include the following, which are discussed further by Aitchison (1983)

(i) It was noted above that the constraint (11.3.1) introduces a negative bias to the correlations between the elements of $\mathbf{x}$, so that 'independence' between variables does not imply zero correlations. A number of ideas have been put forward concerning what should constitute 'independence', and what 'null correlations' are implied, for compositional data. Aitchison (1982) presents arguments in favour of a definition of independence in terms of the structure of the covariance matrix of $\mathbf{v}^{(j)}$—see his equations (4.1) and (5.1). With this definition, the PCs based on $\mathbf{v}$ (or $\mathbf{v}^{(j)}$) for a set of 'independent' variables will simply be the elements of $\mathbf{v}$ (or $\mathbf{v}^{(j)}$), arranged in descending size of their variances. This is equivalent to what happens in PCA for 'ordinary' data with independent variables.

(ii) There is a tractable class of probability distributions for $\mathbf{v}^{(j)}$, and for linear contrasts of the elements of $\mathbf{v}^{(j)}$, but there is no such tractable class for linear contrasts of the elements of $\mathbf{x}$, when $\mathbf{x}$ is restricted by the constraint (11.3.1).

(iii) Because the log-ratio transformation removes the effect of the constraint on the interpretation of covariance, it is possible to define distances between separate observations of $\mathbf{v}$, in a way that is not possible with $\mathbf{x}$.

(iv) It is easier to examine the variability of subcompositions (subsets of $\mathbf{x}$ renormalized to sum to unity), compared to that of the whole composition, if the comparison is done in terms of $\mathbf{v}$ rather than $\mathbf{x}$.

Finally, Aitchison (1983) provides examples in which the proposed PCA of $\mathbf{v}$ is considerably superior to a PCA of $\mathbf{x}$. This seems to be chiefly because there is curvature inherent in many compositional data sets; the proposed analysis is very successful in uncovering correct curved axes of maximum variation, whereas the usual PCA, which is restricted to linear functions of $\mathbf{x}$, cannot. However, Aitchison's proposal does not necessarily make much difference to the results of a PCA, as is illustrated in the following example.

11.3.1. Example: 100 km Running Data

In Section 5.3 a data set was discussed, which consisted of times taken, for each of ten 10 km sections, by 80 competitors in a 100 km race. If, instead of recording the actual time taken in each section, we look at the proportion of the total time taken for each section, the data then becomes compositional in nature. A PCA was carried out on these compositional data, and so was a modified analysis as proposed by Aitchison (1983). The coefficients and variances for the first two PCs are given for the unmodified and modified analyses in Tables 11.1, 11.2, respectively. It can be seen that the PCs in Tables 11.1 and 11.2 are very similar, and this similarity continues with later PCs. The first PC is essentially a linear contrast between times early and late in the race, whereas the second PC is a 'quadratic' contrast, with times early and late in the race contrasted with those in the middle.

Comparison of Tables 11.1. and 11.2 with Table 5.2, shows that converting the data to compositional form has removed the first (overall time) component, but the second PC in Table 5.2 is very similar to the first PC in Tables 11.1 and 11.2. This correspondence continues to later PCs, with the third, fourth, ... PCs for the 'raw' data being very similar to the second, third, ... PCs for the compositional data.

Table 11.1. First two PCs: 100 km compositional data.

	Component 1	Component 2
First 10 km	0.42	0.19
Second 10 km	0.44	0.18
Third 10 km	0.44	0.00
Fourth 10 km	0.40	−0.23
Fifth 10 km	0.05	−0.56
Sixth 10 km	−0.18	−0.53
Seventh 10 km	−0.20	−0.15
Eighth 10 km	−0.27	−0.07
Ninth 10 km	−0.24	0.30
Tenth 10 km	−0.27	0.41
Eigenvalue	4.30	2.31
Cumulative percentage of total variation	43.0	66.1

(First 10 km through Tenth 10 km braced as "Coefficients")

Table 11.2. First two PCs: Aitchison's (1983) technique for 100 km compositional data.

	Component 1	Component 2
First 10 km	0.41	0.19
Second 10 km	0.44	0.17
Third 10 km	0.42	−0.06
Fourth 10 km	0.36	−0.31
Fifth 10 km	−0.04	−0.57
Sixth 10 km	−0.25	−0.48
Seventh 10 km	−0.24	−0.08
Eighth 10 km	−0.30	−0.01
Ninth 10 km	−0.24	0.30
Tenth 10 km	−0.25	0.43
Eigenvalue	4.38	2.29
Cumulative percentage of total variation	43.8	66.6

(First 10 km through Tenth 10 km braced as "Coefficients")

11.4. Principal Component Analysis in Designed Experiments

In Chapters 8 and 9, we discussed ways in which PCA could be used as a preliminary to, or in conjunction with, other standard statistical techniques. The present section gives another example of the same type of application; here we consider the situation where p variables are measured in the course of a designed experiment. The standard analysis would be either a set of separate analyses of variance (ANOVAs) for each variable or, if the variables are correlated, a multivariate analysis of variance (MANOVA) could be performed.

As an illustration, consider a two-way model of the form

$$x_{ijk} = \mu + \tau_j + \beta_k + \varepsilon_{ijk}, \qquad i = 1, 2, \ldots, n_{jk}; j = 1, 2, \ldots, t; k = 1, 2, \ldots, b;$$

where x_{ijk} is the ith observation for treatment j, in block k, of a p-variate vector x. x_{ijk} is therefore the sum of an overall mean μ, a treatment effect τ_j, a block effect β_k, and an error term ε_{ijk}.

The simplest way in which PCA can be used in such analyses is to simply replace the original p variables by their PCs. This sort of replacement is advocated by Rao (1958) in the context of analysing growth curves, but, as noted by Rao (1964), for most types of designed experiment, this simple analysis will often not be particularly useful. This is because the covariance matrix will represent a *mixture* of contributions from within treatments and blocks, between treatments, between blocks, and so on, whereas we will usually wish to *separate* these various types of covariance. This is a more complicated manifestation of what occurs in discriminant analysis (Section 9.1), where a PCA based on the covariance matrix of the raw data may prove confusing, since it inextricably mixes up variation between and within populations. Instead of a PCA of all the x_{ijk}, a number of other PCAs have been suggested and found to be useful in some circumstances.

Jeffers (1962) looks at a PCA of the (treatment × block) means $\bar{x}_{jk}$, $j = 1, 2, \ldots, t, k = 1, 2, \ldots, b$, where

$$\bar{x}_{jk} = \frac{1}{n_{jk}} \sum_{i=1}^{n_{jk}} x_{ijk}$$

(i.e. a PCA of a data set with tb observations on a p-variate random vector). In an example on tree seedlings, he finds that ANOVAs carried out on the first five PCS, which account for over 97% of the variation in the original eight variables, give significant differences between treatment means (averaged over blocks) for the first and fifth PCs. This result contrasts with ANOVAs for the original variables where there were no significant differences. The first and fifth PCs can be readily interpreted in Jeffers' (1962) example, so that transforming to PCs produces a clear advantage in terms of detecting interpretable treatment differences. However, PCs will not always be interpretable, and, as in regression (Section 8.2), there is no reason to expect that treatment differences will manifest themselves in high-variance, rather than low-variance, PCs.

Jeffers (1962) looked at 'between' treatments and blocks PCs, but the PCs of the 'within' treatments, or blocks, covariance matrices can also provide useful information, and the SPSS^X computer package—see Appendix A2.5— has an option for finding such 'error covariance matrix PCs'. Pearce and Holland (1960) give an example with four variables, in which different treatments correspond to different rootstocks, but which has no block structure. They carry out separate PCAs for within- and between-rootstock variation. The first PC is similar in the two cases, measuring general size. Later PCs are, however, different for the two analyses, but they are readily interpretable in both cases, so that the two analyses each provide useful, but separate, information.

Another use of 'within-treatments' PCs occurs in the case where there are several populations, as in discriminant analysis (see Section 9.1), and 'treatments' are defined to correspond to different populations. If each population has the same covariance matrix Σ, and within-population PCs based on Σ are of interest, then the 'within-treatments' covariance matrix provides an estimate of Σ. Yet another way in which 'error covariance matrix PCs' could be used is, as suggested for multivariate regression in Section 10.1, to look for potential outliers.

A different way of using PCA in a designed experiment is described by Mandel (1971, 1972). He considers a situation where there is only one variable, which follows the two-way model

$$x_{jk} = \mu + \tau_j + \beta_k + \varepsilon_{jk}, \qquad j = 1, 2, \ldots, t; k = 1, 2, \ldots, b, \quad (11.4.1)$$

i.e. there is only a single observation on the variable x at each combination of treatments and blocks. In Mandel's analysis, estimates $\hat{\mu}$, $\hat{\tau}_j$, $\hat{\beta}_k$ are found for μ, τ_j, β_k, and residuals are calculated as $e_{jk} = x_{jk} - \hat{\mu} - \hat{\tau}_j - \hat{\beta}_k$. The main interest is then in using e_{jk} to estimate the non-additive part, ε_{jk}, of the model (11.4.1). This non-additive part is assumed to take the form

$$\varepsilon_{jk} = \sum_{h=1}^{m} u_{jh} l_h a_{kh}, \qquad (11.4.2)$$

where m, l_h, u_{jh}, a_{kh} are suitably chosen constants. Apart from slight changes in notation, the right-hand side of (11.4.2) is the same as that of the SVD in (3.5.3). Thus, the model (11.4.2) is fitted by finding the SVD of the matrix $\mathbf{E}$ whose (j, k)th element is e_{jk}, or equivalently finding the PCs of the covariance matrix based on the data matrix $\mathbf{E}$. This analysis also has links with correspondence analysis—see Section 11.1. In both cases we find a SVD of a two-way table of residuals, the difference being that, in the present case, the elements of the table are residuals from an additive model for a quantitative variable, rather than residuals from an independence (multiplicative) model for counts. Freeman (1975) shows that Mandel's approach can be used for incomplete, as well as complete, two-way tables, and illustrates the idea with examples.

A final comment in the present section is that there is an intriguing similarity between PCA and some aspects of the theory of optimal design, which does not seem to have been noted before. Suppose that, as in Chapter 8, we have the linear model

$$E(\mathbf{y}) = \mathbf{X}\boldsymbol{\beta},$$

where $\mathbf{X}$ is the $(n \times p)$ design matrix whose ith row consists of the values of p variables (some of which may be dummy variables) for the ith observation, $i = 1, 2, \ldots, n$, $\boldsymbol{\beta}$ is a vector of p unknown parameters, and $\mathbf{y}$ is a $(n \times 1)$ vector whose ith element is the value of the dependent variable, y, for the ith observation. In optimal design, we wish to choose $\mathbf{X}$ so as to minimize the variances of the elements of $\hat{\boldsymbol{\beta}}$, the least squares estimator for $\boldsymbol{\beta}$. Because we are interested simultaneously in several variances, there are a number of ways in which we can construct a single criterion to optimize. Probably the best-known criterion is to maximize $\det(\mathbf{X}'\mathbf{X})$, which leads to D-optimal designs (Silvey, 1980, p. 10). The criterion is justified by noting that the covariance matrix for $\hat{\boldsymbol{\beta}}$ is proportional to the matrix $(\mathbf{X}'\mathbf{X})^{-1}$. Minimizing the volume of confidence ellipsoids for $\boldsymbol{\beta}$, based on $\hat{\boldsymbol{\beta}}$, leads to minimizing the generalized variance of $\hat{\boldsymbol{\beta}}$, i.e. minimizing $\det[(\mathbf{X}'\mathbf{X})^{-1}]$ (see the discussion following Property A4 in Section 2.1, and Silvey (1980. p. 10)) or, equivalently, maximizing $\det(\mathbf{X}'\mathbf{X})$. It turns out that a design is D-optimal if and only if the quantity $\mathbf{z}((\mathbf{X}'\mathbf{X})/n)^{-1}\mathbf{z}$ is $\leq p$, for all p-element vectors $\mathbf{z}$, and the maximum value, p, is achieved for this quadratic form whenever $\mathbf{z}$ is a row of the optimal design matrix $\mathbf{X}$ (see Fedorov, 1972, Theorem 2.2.1 and its corollary). To see the possible connection with PCA, note that $\mathbf{X}'\mathbf{X}/n$ is a covariance matrix if $\mathbf{X}$ is considered as an $(n \times p)$ data matrix. $\mathbf{X}'\mathbf{X}/n$ and its inverse have the same eigenvectors, and because the rows of $\mathbf{X}$ maximize the quadratic form above, they each must be eigenvectors of $\mathbf{X}'\mathbf{X}/n$. Thus in D-optimal design, we are searching for a design matrix which, when thought of as a data matrix, has PCs whose variances are all equal, and whose coefficients appear as the rows of the data matrix.

11.5. Common Principal Components in the Presence of Group Structure and Comparisons of Principal Components

Suppose that observations on a p-variate random vector $\mathbf{x}$ may have come from any one of g distinct populations, and that the mean and covariance matrix for the ith population are, respectively, $\boldsymbol{\mu}_i$, $\boldsymbol{\Sigma}_i$, $i = 1, 2, \ldots, g$. This is the situation found in discriminant analysis—see Section 9.1—although in discriminant analysis it is often assumed that all the $\boldsymbol{\Sigma}_i$ are the same, so that the populations differ only in their means. If the $\boldsymbol{\Sigma}_i$ are all the same then the 'within-population PCs' are the same for all g populations, though, as pointed out in Section 9.1, within-population PCs are often different from PCs found by pooling data from all populations together.

If the $\boldsymbol{\Sigma}_i$ are different, then there is no uniquely-defined set of within-population PCs which is common to all populations. However, Flury (1984) and Krzanowski (1984a) have examined 'common principal components' which can usefully be defined in some circumstances where the $\boldsymbol{\Sigma}_i$ are not all equal. The idea of 'common' PCs arises if we suspect that the same components underlie the covariance matrices of each group, but that they have different weights in different groups. For example, if anatomical measurements are made on different but closely-related species of animals, then the same general 'size' and 'shape' components (see Section 4.1) may be present for each species, but with varying importance. Similarly, if the same variables are measured on the same individuals but at different times (i.e. groups correspond to different times), then the components may remain the same but their relative importance may vary with time.

The presence of 'common PCs', as just defined, may be expressed formally by the hypothesis that there is an orthogonal matrix $\mathbf{A}$ which simultaneously diagonalizes all the $\boldsymbol{\Sigma}_i$, i.e.

$$\mathbf{A}'\boldsymbol{\Sigma}_i\mathbf{A} = \boldsymbol{\Lambda}_i, \tag{11.5.1}$$

where $\boldsymbol{\Lambda}_i$, $i = 1, 2, \ldots, g$ are all diagonal. The kth column of $\mathbf{A}$ gives the coefficients of the kth common PC, and the (diagonal) elements of $\boldsymbol{\Lambda}_i$ give the variances of these PCs for the ith population. Note that the order of these variances need not be the same for all i, so that different PCs may have the largest variance in different populations.

Flury (1984) discusses how to test the hypothesis implied by (11.5.1), and how to find maximum likelihood estimates of $\mathbf{A}$ and $\boldsymbol{\Lambda}_i$, in both cases assuming multivariate normality. The hypothesis (11.5.1) has been generalized by Flury (1985), to allow a subset of q $(<p)$ PCs to be common to all groups. The same paper also considers the possibility that q PCs span the same q-dimensional subspace in each group, without requiring individual PCs to be identical.

Krzanowski (1984a) describes a simpler method of obtaining estimates of $\mathbf{A}$ and $\boldsymbol{\Lambda}_i$, based on the fact that if (11.5.1) is true then the columns of $\mathbf{A}$ contain the eigenvectors not only of $\boldsymbol{\Sigma}_1, \boldsymbol{\Sigma}_2, \ldots, \boldsymbol{\Sigma}_g$ individually but of any linear combination of $\boldsymbol{\Sigma}_1, \boldsymbol{\Sigma}_2, \ldots, \boldsymbol{\Sigma}_g$. He therefore uses the eigenvectors of $\mathbf{S}_1 + \mathbf{S}_2 + \cdots + \mathbf{S}_g$, where $\mathbf{S}_i$ is the sample covariance matrix for the ith population, to estimate $\mathbf{A}$, and then substitutes this estimate and $\mathbf{S}_i$, for $\mathbf{A}$ and $\boldsymbol{\Sigma}_i$ respectively, in (11.5.1) to obtain estimates of $\boldsymbol{\Lambda}_i$, $i = 1, 2, \ldots, g$.

To assess whether or not (11.5.1) is true, the estimated eigenvectors of $\mathbf{S}_1 + \mathbf{S}_2 + \cdots + \mathbf{S}_g$ can be compared with those estimated for some other weighted sum of $\mathbf{S}_1, \mathbf{S}_2, \ldots, \mathbf{S}_g$, chosen to have different eigenvectors from $\mathbf{S}_1 + \mathbf{S}_2 + \cdots + \mathbf{S}_g$ if (11.5.1) does not hold. The comparison between eigenvectors can be made either informally, or using methodology developed by Krzanowski (1979b), which is now described.

Suppose that sample covariance matrices $\mathbf{S}_1, \mathbf{S}_2$ are available for two groups of individuals, and we wish to compare the two sets of PCs found from $\mathbf{S}_1$ and $\mathbf{S}_2$. Let $\mathbf{A}_{1q}, \mathbf{A}_{2q}$ be $(p \times q)$ matrices whose columns contain the coefficients of the first q PCs based on $\mathbf{S}_1, \mathbf{S}_2$ respectively. Krzanowski's (1979b) idea is to find the minimum angle, δ, between the subspaces defined by the q columns of $\mathbf{A}_{1q}$ and $\mathbf{A}_{2q}$, together with associated vectors in these two subspaces which subtend this minimum angle. It turns out that δ is given by

$$\delta = \cos^{-1}(v_1^{1/2}),$$

where v_1 is the largest eigenvalue of $\mathbf{A}'_{1q}\mathbf{A}_{2q}\mathbf{A}'_{2q}\mathbf{A}_{1q}$, and the vectors which subtend the minimum angle are related, in a simple way, to the corresponding eigenvector.

The analysis can be extended by looking at the second, third, ... eigenvalues and corresponding eigenvectors of $\mathbf{A}'_{1q}\mathbf{A}_{2q}\mathbf{A}'_{2q}\mathbf{A}_{1q}$; from these can be found pairs of vectors, one each in the two subspaces spanned by $\mathbf{A}_{1q}$, $\mathbf{A}_{2q}$, which have minimum angles between them, subject to being orthogonal to previous pairs of vectors. Another extension is to $g > 2$ groups of individuals with covariance matrices $\mathbf{S}_1, \mathbf{S}_2, \ldots, \mathbf{S}_g$, and matrices $\mathbf{A}_{1q}, \mathbf{A}_{2q}, \ldots, \mathbf{A}_{gq}$, containing the first q eigenvectors (PC coefficients) for each group. We can then look for a vector, which minimizes

$$\Delta = \sum_{i=1}^{g} \cos^2 \delta_i,$$

where δ_i is the angle which the vector makes with the subspace defined by the columns of $\mathbf{A}_{iq}$. This objective is achieved by finding eigenvalues and eigenvectors of

$$\sum_{i=1}^{g} \mathbf{A}_{iq}\mathbf{A}'_{iq},$$

and for $g = 2$, the analysis reduces to that given above.

In Krzanowski (1979b) the technique was suggested as a descriptive technique—if δ (for $g = 2$) or Δ (for $g > 2$) is 'small enough' then the subsets of q PCs for the g groups are similar, but there is no formal definition of 'small enough'. In a later paper, Krzanowski (1982) investigates the behaviour of δ using simulation, both when all the individuals come from populations with the same covariance matrix, and when the covariance matrices are different for the two groups of individuals. The simulation encompasses several different values for p, q, and for the sample sizes, and it also includes several different structures for the covariance matrices (though it does not look at correlation matrices). Krzanowski (1982) is therefore able to offer some limited guidance on what constitutes a 'small enough' value of δ, based on the results from his simulations.

As an example, anatomical data similar to that discussed in Sections 4.1, 5.1, 9.2, 10.1 and 10.2 has been collected for different groups of students in different years. Comparing the first three PCs found for the 1982 and 1983 groups of students gives a value of $2.02°$ for δ; the corresponding value for 1982 and 1984 is $1.25°$, and that for 1983 and 1984 is $0.52°$. Krzanowski (1982) does not have a table of simulated critical angles for the sample sizes and number of variables relevant to this example. However, the three values quoted above are well below the critical values for $p = 8$ and sample sizes of 50. In the present example, all sample sizes are approximately 30 (with $p = 7$), so that the critical values will be larger than for samples of size 50—see Krzanowski (1982, Table I). Hence, as might be expected from such small angles, the sets of the first three PCs are not significantly different for the three years 1982, 1983, 1984.

If all three years are compared simultaneously, then the angles between the subspaces formed by the first three PCs, and the nearest vector to all three subspaces are:

1982	1983	1984
1.17°	1.25°	0.67°

Again, the angles are very small; although no tables are available for assessing the significance of these angles, they seem to confirm the impression given by looking at the years two at a time, that the sets of the first three PCs are not significantly different for the three years.

Two points should be noted with respect to Krzanowski's technique. First, it can only be used to compare *subsets* of PCs—if $q = p$, then $\mathbf{A}_{1p}$, $\mathbf{A}_{2p}$ will usually span p-dimensional space (unless either $\mathbf{S}_1$ or $\mathbf{S}_2$ has zero eigenvalues), so that δ is necessarily zero. It seems likely that the technique will be most valuable for values of q which are small compared to p. The second point is that while δ is clearly a useful measure of the closeness of two subsets of PCs, the vectors and angles found from the second, third, …, eigenvalues and eigenvectors of $\mathbf{A}'_{1q}\mathbf{A}_{2q}\mathbf{A}'_{2q}\mathbf{A}_{1q}$ are, successively, less valuable. The first two or three angles give an idea of the overall difference between the two subspaces, provided that q is not too small. However, if we reverse the analysis and look at the *smallest* eigenvalue, and corresponding eigenvector, of $\mathbf{A}'_{1q}\mathbf{A}_{2q}\mathbf{A}'_{2q}\mathbf{A}_{1q}$, then we find the *maximum* angle between vectors in the two subspaces (which will often be 90°, unless q is small). Thus, the last few angles, and corresponding vectors, will need to be interpreted in a rather different way from that of the first few. The general problem of interpreting angles other than the first can be illustrated by again considering the first three PCs for the student anatomical data from 1982 and 1983. We saw above that $\delta = 2.02°$, which is clearly very small; the second and third angles for these data are 25.2° and 83.0°. These angles are fairly close to the 5% critical values given in Krzanowski (1982) for the second and third angles when $p = 8$ and the sample sizes are each 50, but it is difficult to see what this result implies. In particular, the fact that the third angle is close to 90° might intuitively suggest that the first three PCs are significantly different for 1982 and 1983. Intuition is, however, contradicted by Krzanowski's Table I, which shows that for sample sizes as small as 50 (and, hence, certainly for samples of size 30), the 5% critical value for the third angle is nearly 90°.

Muller (1982) suggests that canonical correlation analysis of PCs (see Section 9.3) provides a way of comparing the PCs based on two sets of variables, and cites some earlier references. When the two sets of variables are, in fact, the same variables measured for two groups of observations, Muller's analysis is equivalent to that of Krzanowski; the latter author pointed out links between canonical correlation analysis and his own technique in his 1979 paper.

Preisendorfer and Mobley (1982), in a series of five technical reports, have examined various ways of comparing data sets, measured on the same variables at different times, and part of their work involves comparison of PCs from different sets—see, in particular, their third report, which concentrates on comparing the SVDs of two data matrices $\mathbf{X}_1$, $\mathbf{X}_2$, say. Suppose that the SVDs are written

$$\mathbf{X}_1 = \mathbf{U}_1 \mathbf{L}_1 \mathbf{A}_1', \quad \text{and}$$

$$\mathbf{X}_2 = \mathbf{U}_2 \mathbf{L}_2 \mathbf{A}_2'.$$

Then Preisendorfer and Mobley (1982) define a number of statistics which compare $\mathbf{U}_1$ with $\mathbf{U}_2$, $\mathbf{A}_1$ with $\mathbf{A}_2$, $\mathbf{L}_1$ with $\mathbf{L}_2$, or compare any two of the three factors in the SVD for $\mathbf{X}_1$ with the corresponding factors in the SVD for $\mathbf{X}_2$. All of these comparisons are relevant to comparing PCs, since $\mathbf{A}$ contains the coefficients of the PCs, $\mathbf{L}$ provides the standard deviations of the PCs, and $\mathbf{U}$ gives PC scores, normalized to have variance $1/(n-1)$ (see Section 3.5). The 'significance' of an observed value of any one of Preisendorfer and Mobley's statistics is assessed by comparing the value to a 'reference distribution', which is obtained by simulation.

11.6. Principal Component Analysis in the Presence of Missing Data

In all the examples given in this text, the data set has been complete. However, it is not uncommon, especially for large data sets, for some of the values of some of the variables to be missing. The most usual way of dealing with such a situation is to delete, entirely, any observation for which at least one of the variables has a missing value. This is satisfactory if missing values are few, but clearly wasteful of information if a high proportion of observations have missing values for just one or two variables. To meet this problem, a number of alternatives have been suggested, and one of the computer packages discussed in the Appendix (SPSS^X—see Appendix A.2.5) has options for two such alternatives, which involve different ways of computing the covariance or correlation matrix.

The first option is to compute the (j, k)th correlation or covariance using all observations for which values of both x_j and x_k are available. Unfortunately, this leads to covariance or correlation matrices which are not necessarily positive semidefinite. The second option in SPSS^X is to replace missing values for variable x_j by the mean value, $\bar{x}_j$, on the observations for which the value of x_j is available—according to Beale and Little (1975), this approach is fairly common and has produced satisfactory results. Beale and Little's (1975) paper deals mainly with multiple regression, but some of their discussion is in terms of missing values in a more general multivariate context. The same authors also note a modification of the first SPSS^X option, in which $\bar{x}_j$, $\bar{x}_k$ are calculated from all available values of x_j, x_k, respectively, instead of only for observations for which both x_j and x_k have values present, when computing the summation $\sum_i (x_{ij} - \bar{x}_j)(x_{ik} - \bar{x}_k)$ in the covariance or correlation matrix. They state that, at least in the regression context, the results can be unsatisfactory.

The main method discussed by Beale and Little (1975) is based on the assumption of multivariate normality. The covariances and correlations are estimated, under this assumption, using the method of maximum likelihood. Maximum likelihood estimation of the covariance matrix under the same conditions has been considered by a number of other authors, for example Anderson (1957), and if multivariate normality is a valid assumption, then this method is entirely appropriate. However, it is not clear how satisfactory the technique will be, if multivariate normality does not hold. Remaining with the theme of maximum likelihood estimation, De Ligny et al. (1981) describe an algorithm for factor analysis (by which they apparently mean PCA) with missing data. They claim that their algorithm will produce maximum likelihood estimates, but concede that other procedures may exist with better computational properties.

Wiberg (1976) produced one of very few published papers which deal specifically with PCA when discussing missing data. His approach is via the SVD, which gives a least squares approximation of rank m to the data matrix $\mathbf{X}$. In other words, the approximation $_m\tilde{x}_{ij}$ minimizes

$$\sum_{i=1}^{n} \sum_{j=1}^{p} (_m x_{ij} - x_{ij})^2,$$

where $_m x_{ij}$ is any rank m approximation to x_{ij}—see Section 3.5. Principal components can be computed from the SVD (see Section 3.5 and Appendix A1). With missing data, Wiberg (1976) proposes that we minimize the same quantity, but with the summation only over values of (i, j) for which x_{ij} is not missing; PCs can then be estimated from the modified SVD. The same idea is implicitly suggested by Gabriel and Zamir (1979). Wiberg (1976) reports that for simulated multivariate normal data his method is slightly worse than the method based on maximum likelihood estimation. However, his method has the virtue that it can be used regardless of whether or not the data come from a multivariate normal distribution.

For the specialized use of PCA in analysing residuals from an additive model, for data from designed experiments (see Section 11.4), Freeman (1975) has shown that incomplete data can be easily handled.

Finally, note that there is a similarity of purpose in robust estimation of PCs (see Section 10.3) to that present in handling missing data. In both cases we identify particular observations which we cannot use in unadjusted form, either because they are suspiciously extreme (in robust estimation), or because they are not given at all (missing values). To completely ignore such observations may throw away valuable information, so we can try to estimate 'correct' values for the observations in question. Similar techniques could be tried in each case. For example, one of the methods described in Section 10.3 involved regression of the variables on each other, and, in the missing data context, Frane (1976) has suggested estimating missing values for a particular observation by means of regression analyses for the missing variables, on the variables that are present for the given observation.

11.7. Principal Components for Goodness-of-Fit Statistics

This final section discusses briefly an application of PCA which has a number of unusual features, and which does not fit very naturally into any of the chapters of this book. The context of the application is testing whether or not a (univariate) set of data $y_1, y_2, \ldots, y_n$ could have arisen from a given probability distribution, with cumulative distribution function $G(y)$, i.e. we want a goodness-of-fit test. If the transformation

$$x_i = G(y_i), \qquad i = 1, 2, \ldots, n,$$

is made, then we can, equivalently, test whether or not $x_1, x_2, \ldots, x_n$ are from a uniform distribution on the range (0, 1). Assume, without loss of generality, that $x_1 \le x_2 \le \cdots \le x_n$, and define the sample distribution function as $F_n(x) = i/n$ for $x_i \le x < x_{i+1}$, $i = 0, 1, \ldots, n$, where x_0, x_{n+1} are defined as 0, 1 respectively. Then a well-known test statistic is the Cramér-von Mises statistic

$$W_n^2 = n \int_0^1 (F_n(x) - x)^2 \, dx.$$

Like most all-purpose goodness-of-fit statistics, W_n^2 will detect many different types of discrepancy between the observations and $G(y)$, and a large value of W_n^2 on its own will give no information about what type of discrepancy has occurred. For this reason a number of authors (see, for example, Durbin and Knott, 1972; Durbin et al., 1975) have looked at decompositions of W_n into a number of separate 'components', each of which measures the degree to which a different type of discrepancy is present.

It turns out that a 'natural' way of partitioning (Durbin and Knott, 1972) is

$$W_n^2 = \sum_{k=1}^{\infty} \frac{z_{nk}^2}{k^2 \pi^2},$$

where

$$z_{nk} = (2n)^{1/2} k\pi \int_0^1 (F_n(x) - x) \sin(k\pi x) \, dx, \qquad k = 1, 2, \ldots,$$

are PCs of $\sqrt{n}(F_n(x) - x)$. The phrase 'PCs of $\sqrt{n}(F_n(x) - x)$' needs further explanation, since $\sqrt{n}(F_n(x) - x)$ is not, as is usual when defining PCs, a p-variable vector. Instead, it is an infinite-dimensional random variable corresponding to the continuum of values for x between 0 and 1. The covariance matrix for $\sqrt{n}(F_n(x) - x)$ therefore needs to be defined in a rather different manner from usual. However, with an appropriate definition of the covariance matrix, eigenvectors, and corresponding PCs, it follows that z_{nk}, $k = 1, 2, \ldots$, as defined above, are, indeed, PCs for $\sqrt{n}(F_n(x) - x)$.

The components z_{nk}, $k = 1, 2, \ldots$ are discussed in considerable detail, from both theoretical and practical viewpoints, by Durbin and Knott (1972), and Durbin et al. (1975), who also give several additional references for the topic.

CHAPTER 12

Generalizations and Adaptations of Principal Component Analysis

The basic technique of PCA has been generalized or adapted in many ways, and some have already been discussed, in particular in Chapter 11 where adaptations for special types of data were described. This final chapter discusses a number of additional generalizations and modifications; for several of them the discussion is very brief in comparison to the large amount of material that has appeared in the literature, because most have yet to be used widely in practice.

Sections 12.1 and 12.2 present, respectively, some definitions of 'generalized PCA' (which includes 'weighted PCA') and 'non-linear PCA'. In both cases there are connections with correspondence analysis, which was discussed at somewhat greater length in Section 11.1. Section 12.3 discusses procedures for finding 'PCs' based on data matrices which are uncentred or doubly-centred rather than, as is usual, centred by columns but not rows.

Section 12.4 discusses both the possibility of simplifying PCs by restricting their coefficients to a small number of distinct values, and the related topic of the sensitivity of PC coefficients to changes in the variances of the PCs. Section 12.5 describes modifications of PCA which may be useful when 'instrumental variables' are present.

In Section 12.6 some possible alternatives to PCA when the data come from known non-normal distributions are discussed, and in Section 12.7 the ideas of three-mode PCA, are introduced. This latter type of analysis is appropriate when the data matrix, as well as having two dimensions corresponding to individuals and variables respectively, has a third dimension corresponding, for example, to time.

A final section, Section 12.8, presents a few concluding remarks.

12.1. Generalized and Weighted Principal Component Analysis

Both Greenacre (1984, Appendix A) and Gnanadesikan (1977, Section 2.4.2) discuss procedures which they call 'generalized PCA'. The two topics are different and in the latter case, where the idea is to allow non-linear functions of the elements of x in the PCA, there are connections with the next section, on non-linear PCA.

Consider first Greenacre's (1984) definition, and recall the singular value decomposition (SVD) of the $(n \times p)$ data matrix X, defined in equation (3.5.1), namely

$$X = ULA'. \tag{12.1.1}$$

Then the matrices A, L give, respectively, the eigenvectors and the square roots of the eigenvalues of $X'X$, and hence the coefficients and variances of the PCs for the sample covariance matrix S.

In equation (12.1.1) we have $U'U = I_r$, $A'A = I_r$, where r is the rank of X, and I_r is the identity matrix of order r. Suppose now that Ω and Φ are specified positive-definite symmetric matrices and that we replace (12.1.1) by a 'generalized SVD',

$$X = VMB', \tag{12.1.2}$$

where V, B are $(n \times r)$, $(p \times r)$ matrices respectively, satisfying $V'\Omega V = I_r$, $B'\Phi B = I_r$, and M is a $(r \times r)$ diagonal matrix.

This representation follows by finding the ordinary SVD of $\tilde{X} = \Omega^{1/2} X \Phi^{1/2}$. If we write the usual SVD of $\tilde{X}$ as

$$\tilde{X} = WKC', \tag{12.1.3}$$

where K is diagonal, $W'W = I_r$, $C'C = I_r$, then

$$X = \Omega^{-1/2} \tilde{X} \Phi^{-1/2}$$

$$= \Omega^{-1/2} WKC' \Phi^{-1/2}.$$

Putting $V = \Omega^{-1/2} W$, $M = K$, $B = \Phi^{-1/2} C$ gives (12.1.2), where M is diagonal, $V'\Omega V = I_r$, and $B'\Phi B = I_r$, as required. With this representation, Greenacre (1984) defines generalized PCs as having coefficients given by the columns of B, in the case where Ω is diagonal. Rao (1964) suggested a similar modification of PCA, to be used when oblique, rather than orthogonal, axes are desired. His idea is to use the transformation $Z = XB$, where $B'\Phi B = I$, for some specified positive-definite matrix, Φ; this idea clearly has links with generalized PCs, as just defined.

It was noted in Section 3.5 that, if we take the usual SVD and retain only the first m PCs, so that x_{ij} is approximated by

$$_m\tilde{x}_{ij} = \sum_{k=1}^{m} u_{ik} l_k^{1/2} a_{jk} \quad \text{(with notation as in Section 3.5),}$$

then $_m\tilde{x}_{ij}$ provides a best possible rank m approximation to x_{ij} in the sense of minimizing $\sum_{i=1}^{n}\sum_{j=1}^{p}(_mx_{ij} - x_{ij})^2$ among all possible rank m approximations $_mx_{ij}$. It can also be shown (Greenacre, 1984, p. 39) that if $\boldsymbol{\Omega}$, $\boldsymbol{\Phi}$ are both diagonal matrices, with elements ω_i, $i = 1, 2, \ldots, n$, ϕ_j, $j = 1, 2, \ldots, p$, respectively, and if $_m\tilde{\tilde{x}}_{ij} = \sum_{k=1}^{m} v_{ik}m_k b_{jk}$, where v_{ik}, m_k, b_{jk} are elements of $\mathbf{V}$, $\mathbf{M}$, $\mathbf{B}$ defined in (12.1.2), then $_m\tilde{\tilde{x}}_{ij}$ minimizes

$$\sum_{i=1}^{n}\sum_{j=1}^{p} \omega_i \phi_j (_mx_{ij} - x_{ij})^2, \tag{12.1.4}$$

among all possible rank m approximations $_mx_{ij}$ to x_{ij}. Thus, the special case of generalized PCA, in which $\boldsymbol{\Phi}$ as well as $\boldsymbol{\Omega}$ is diagonal, is a form of 'weighted PCA', where different variables can have different weights, ϕ_1, $\phi_2, \ldots, \phi_p$, and different observations can have different weights $\omega_1, \omega_2, \ldots,$ ω_n. Cochran and Horne (1977) discuss the use of this type of weighted PCA in a chemical context.

It is possible, of course, that one set of weights, but not the other, is present. For example, if $\omega_1 = \omega_2 = \cdots = \omega_n$, but the ϕ's may be different, then only the variables have different weights and the observations are treated identically. Using the correlation matrix, rather than the covariance matrix, is a special case, in which $\phi_j = 1/s_{jj}$ where s_{jj} is the sample variance of the jth variable. Deville and Malinvaud (1983) argue that the choice of $\phi_j = 1/s_{jj}$ is somewhat arbitrary and that other weights may be appropriate in some circumstances, and Rao (1964) has also suggested the possibility of using weights for variables. Conversely, the computer packages BMDP, SAS and SPSS^X—see Appendix A2—allow different observations to have different weights, although the variables remain equally weighted. In practice, it must be rare that a uniquely appropriate set of ω_i's or ϕ_j's suggests itself, and Greenacre (1984, Table A1) states that the potential of using different ϕ_j's and ω_i's has yet to be explored.

An even more general set of weights than that given in (12.1.4) is proposed by Gabriel and Zamir (1979). Here $\mathbf{X}$ is approximated by minimizing

$$\sum_{i=1}^{n}\sum_{j=1}^{p} w_{ij}(_m\hat{x}_{ij} - x_{ij})^2, \tag{12.1.5}$$

where the rank m approximation to $\mathbf{X}$ has elements $_m\hat{x}_{ij}$ of the form $_m\hat{x}_{ij} = \sum_{k=1}^{m} g_{ik}h_{jk}$ for suitably chosen constants g_{ik}, h_{jk}, $i = 1, 2, \ldots, n, j = 1, 2, \ldots, p$, $k = 1, 2, \ldots, m$. This does not readily fit into the generalized PC framework above unless the w_{ij}'s can be written as products $w_{ij} = \omega_i\phi_j$, $i = 1, 2, \ldots, n$; $j = 1, 2, \ldots, p$, although it involves similar ideas. The examples given by Gabriel and Zamir (1979) can be expressed as contingency tables, so that correspondence analysis rather than PCA may be more appropriate, and Greenacre (1984), too, develops generalized PCA as an offshoot of correspondence analysis (he shows that another special case of the generalized SVD (12.1.2) produces correspondence analysis, a result which was discussed further in Section 11.1). The idea of weighting could, however, be used in PCA for any type of data, provided that suitable weights can be defined.

Gabriel and Zamir (1979) suggest a number of ways in which special cases of their weighted analysis may be used. As noted in Section 11.6, it can accommodate missing data, by giving zero weight to missing elements of **X**.

Alternatively, the analysis can be used to look for 'outlying cells' in a data matrix. This can be achieved by using similar ideas to those introduced in Section 6.1.5 in the context of choosing how many PCs to retain. Any particular element, x_{ij}, of **X** is estimated by least squares based on a subset of the data, which does not include x_{ij}. This (rank m) estimate, $_m\hat{x}_{ij}$, is readily found by equating to zero a subset of weights, including w_{ij}. The difference between x_{ij} and $_m\hat{x}_{ij}$ provides a better measure of how 'outlying' is x_{ij} compared to the remaining elements of **X**, than does the difference between x_{ij} and a rank m estimate, $_m\tilde{x}_{ij}$, based on the SVD for the entire matrix **X**. This result follows because $_m\hat{x}_{ij}$ is not affected by x_{ij}, whereas x_{ij} contributes to the estimate $_m\tilde{x}_{ij}$.

Gnanadesikan's (1977) definition of a 'generalized PCA' is basically to extend the vector of p variables **x**, to include *functions* of the elements of **x**. For example, if $p = 2$, so $\mathbf{x}' = (x_1, x_2)$, we could consider linear functions of $\mathbf{x}'_+ = (x_1, x_2, x_1^2, x_2^2, x_1 x_2)$ which have maximum variance, rather than restricting attention to linear functions of **x**'. In theory, any functions $g_1(x_1, x_2, \ldots, x_p)$, $g_2(x_1, x_2, \ldots, x_p)$, $\ldots$, $g_h(x_1, x_2, \ldots, x_p)$ of $x_1, x_2, \ldots, x_p$ could be added to the original vector **x**, in order to construct an extended vector $\mathbf{x}_+$, whose PCs are then found. In practice, however, Gnanadesikan (1977) concentrates on quadratic functions, so that the analysis is a procedure for finding quadratic, rather than linear, functions of **x** which maximize variance.

The next section gives further information on non-linear varieties of PCA.

12.2. Non-linear Principal Component Analysis

One way of introducing non-linearity into PCA is Gnanadesikan's (1977) 'generalized PCA', which was described in the previous section, but there are a number of alternatives. An obvious one is to *replace* **x** by a function of **x**, rather than *add* to **x** as in Gnanadesikan's analysis. Transforming **x** in this way might be appropriate, for example, if we are interested in products of powers of the elements of **x**. In this case, taking logarithms of the elements and doing a PCA on the transformed data will provide a suitable analysis. Another possible use of transforming to non-linear PCs is in order to detect near-constant non-linear relationships between the variables. If an appropriate transformation is made, such relationships will be detected by the last few PCs of the transformed data. Transforming the data is also suggested, before doing a PCA, for compositional data—see Section 11.3. However, as noted in the introduction to Chapter 4, transformation of variables should only be undertaken, in general, after careful thought about whether it is appropriate for the data set at hand.

De Leeuw (1982) discusses two forms of 'non-linear PCA' and connections between them. A considerable amount of work, mostly rather abstract, has been done by De Leeuw and co-workers on 'non-linear multivariate analysis' but the topic is treated only very briefly here.

In one version of non-linear PCA, termed 'non-metric PCA' by De Leeuw (1982)—see also Section 11.1—a sequence, $L_1, L_2, \ldots$, of closed subspaces of p-dimensional space are searched for components $z_1, z_2, \ldots$, in such a way that the cumulative sums of the largest eigenvalues of a correlation matrix of $z_1, z_2, \ldots$, are maximized. Ordinary PCA is a special case where $L_1, L_2, \ldots$, are one-dimensional and linear. The second version is multiple correspondence analysis (see Greenacre, 1984, Chapter 5), which was also mentioned in Section 11.1. De Leeuw (1982) examines various computational and geometric relationships between the two approaches.

12.3. Non-centred Principal Component Analysis and Doubly-Centred Principal Component Analysis

Principal components are linear functions of $\mathbf{x}$ whose coefficients are given by the eigenvectors of a covariance or correlation matrix, or, equivalently, the eigenvectors of a matrix $\mathbf{X'X}$. Here $\mathbf{X}$ is a $(n \times p)$ matrix whose (i, j)th element is the value, for the ith observation, of the jth variable, *measured about the mean* for that variable. Thus, the columns of $\mathbf{X}$ have been centred, so that the sum of each column is zero.

Two alternatives to 'centring by columns' have been suggested:

(i) the columns of $\mathbf{X}$ are left uncentred, i.e. x_{ij} is now the value, for the ith observation, of the jth variable, *as originally measured*;
(ii) *both* rows and columns of $\mathbf{X}$ could be centred, so that sums of rows, as well as sums of columns, are zero.

In either (i) or (ii) the analysis now proceeds by looking at linear functions of $\mathbf{x}$ whose coefficients are the eigenvectors of $\mathbf{X'X}$, with $\mathbf{X}$ now non-centred or doubly-centred. Of course, these linear functions no longer maximize variance, and so are not PCs according to the usual definition, but it is convenient to refer to them as non-centred and doubly-centred PCs, respectively.

Noncentred PCA seems to be a fairly well-established technique in ecology. It has also been used in chemistry (Cochran and Horne, 1977) and geology (Jöreskog *et al.* 1976, Section 5.2), and it is an option in the BMDP computer package—see Appendix A2.1. As noted by Ter Braak (1983), the technique projects observations onto the best fitting hyperplane *through the origin*, rather than through the centroid of the data set. If the data are such that the origin is an important point of reference, then this type of analysis can be relevant. However, if the centre of the observations is a long way from the origin, then the first 'PC' will dominate the analysis, and will simply reflect the position of the centroid. For data which consist of counts of various biological species (the observations) at various sites (the variables), Ter Braak (1983) claims that non-centred PCA is better than standard (centred) PCA at simultaneously representing within-site diversity *and* between-site diversity of species. Centred PCA is better at representing between-site species diversity than non-centred PCA, but it is more difficult to deduce within-site diversity from a centred PCA.   Jöreskog *et al.* (1976, Section 7.6) discuss an application of the method (which they refer to as Imbrie's Q-mode method) in a similar context, concerning the abundance of various marine micro-organisms in cores taken at various sites on the seabed. The same authors also suggest that this type of analysis is relevant for data where the p variables are amounts of p chemical constituents in n soil or rock samples. If the degree to which two samples have the same *proportions* of each constituent is considered to be an important index of similarity between samples, then the similarity measure implied by non-centred PCA will be appropriate (Jöreskog *et al.*, 1976, p. 89). An alternative approach if proportions are of interest is to reduce the data to compositional form—see Section 11.3.

Doubly-centred PCA has been proposed, by Buckland and Anderson (1985), as another method of analysis for data which consist of species counts at various sites. They argue that centred PCA of such data may be dominated by a 'size' component, which measures the relative abundance of the various species. It is possible to simply ignore the first PC, and concentrate on later PCs, but an alternative is provided by double centring which will 'remove' the 'size' PC. At the same time a PC with zero eigenvalue is introduced, because the constraint $x_{i1} + x_{i2} + \cdots + x_{ip} = 0$ now holds for all i. A further alternative for removing the 'size' effect of different abundances of different species, in some such data sets, would be to record only whether a species is present or absent at each site, rather than the actual counts for each species.

In fact, what is being done in double centring is the same as Mandel's (1971, 1972) approach to data in a two-way analysis of variance—see Section 11.4. We are removing main effects due to rows/observations/species, and due to columns/variables/sites, and concentrating our analysis on the interaction between species and sites. In the regression context, Hoerl *et al.* (1985) have suggested that double centring can remove 'non-essential ill-conditioning', which is caused by the presence of a row (observation) effect in the original data.

One reason for the suggestion of both non-centred and doubly-centred PCA for counts of species at various sites is perhaps that it is not entirely clear which of 'sites' and 'species' should be treated as 'variables' and which as 'observations'. Another possibility would be to centre with respect to species, but not sites, in other words carrying out an analysis with species, rather than sites, as the variables.

Yet another technique which has been suggested for analysing some types of site–species data is correspondence analysis (see, for example, Gauch, 1982). As pointed out in Section 11.4, correspondence analysis has some similarity to Mandel's approach, and hence to doubly-centred PCA. In doubly-centred PCA we analyse the residuals from an additive model for row and column (site and species) effects, whereas in correspondence analysis the residuals from a multiplicative (independence) model are considered.

12.4. Discrete Coefficients for Principal Components and Sensitivity of Principal Components

In Chapter 4, especially its first section, it was mentioned that there is usually no need to express the coefficients of a set of PCs to more than one or two decimal places. This idea could, in theory, be formalized to allow only a finite number of different values of the coefficients, say 0, ± 0.1, ± 0.2, ..., ± 1.0, or, as an extreme, only the values 0 or ± 1. Restricting the coefficients in this way at the outset would lead to a more difficult optimization problem than standard PCA and would give different results; an alternative is simply to do the PCA as usual, and then round the coefficients. The rounded PCs will no longer be exactly orthogonal and their variances will be changed, but, typically, these effects will not be very great, as demonstrated by Green (1977) and Bibby (1980). The latter paper presents bounds on the changes in the variances of the PCs (both in absolute and relative terms) which are induced by rounding coefficients, and shows that in practice the changes are quite small, even with fairly severe rounding.

To illustrate the effect of severe rounding, consider again Table 3.2, in which PCs for eight blood chemistry variables have their coefficients rounded to the nearest 0.2. Thus, the coefficients for the first PC, for example, are given as

$$0.2 \quad 0.4 \quad 0.4 \quad 0.4 \quad -0.4 \quad -0.4 \quad -0.2 \quad -0.2,$$

compared to their values to three decimal places, which are

$$0.195 \quad 0.400 \quad 0.459 \quad 0.430 \quad -0.494 \quad -0.320 \quad -0.177 \quad -0.171.$$

The variance of the simplified PC is 2.536, compared to an exact variance of 2.792, a percentage change of 9%. The angle between the directions defined by the exact and simplified PCs is about 8°. For the second, third and fourth PCs given in Table 3.2, the percentage changes in variances are 7%, 11% and 11%, respectively, and the angles between exact and simplified PCs are 8° in each case. None of these changes and angles are unduly large, considering the severity of the rounding which has been employed.

Bibby (1980) also mentions the possibility of using conveniently chosen integer values for the coefficients. For example, the simplified first PC from Table 3.2 is proportional to $2(x_2 + x_3 + x_4) + x_1 - (x_7 + x_8) - 2(x_5 + x_6)$, which should be much simpler to interpret than the exact form of the PC. In general, this type of simplification will increase the interpretability of individual PCs, but makes comparisons between PCs more difficult, because different PCs will now have different values of $\mathbf{a}_k'\mathbf{a}_k$.

Green (1977) discusses the effect of rounding in PCA in a rather different manner. Instead of investigating the direct effect on the PCs, he looks at the proportions of variance accounted for, *in each individual variable*, by the first m PCs, and examines by how much these proportions are reduced by rounding. He concludes that changes due to rounding are small, even for quite severe rounding, and recommends rounding to the nearest 0.1 or even 0.2, since this will increase interpretability, with little effect on other aspects of the analysis.

Krzanowski (1984b) considers what could be thought of as the opposite problem to that discussed by Bibby (1980). Instead of looking at the effect of small changes in $\boldsymbol{\alpha}_k$ on the value of λ_k, Krzanowski (1984b) examines the effect on $\boldsymbol{\alpha}_k$ of small changes in the value of λ_k. He argues that this is an important problem because it gives information on the stability of PCs; the PCs can only be confidently interpreted if they are stable with respect to small changes in the values of the λ_k's.

If λ_k is *decreased* by an amount ε, then Krzanowski (1984b) looks for a vector $\boldsymbol{\alpha}_k^{\varepsilon}$ which is maximally different from $\boldsymbol{\alpha}_k$, subject to $\mathrm{var}(\boldsymbol{\alpha}_k^{\varepsilon\prime}\mathbf{x}) = \lambda_k - \varepsilon$. He finds that the angle θ between $\boldsymbol{\alpha}_k$ and $\boldsymbol{\alpha}_k^{\varepsilon}$ is given by

$$\cos \theta = [1 + \varepsilon/(\lambda_k - \lambda_{k+1})]^{-1/2}, \qquad (12.4.1)$$

and so depends mainly, on the difference between λ_k, λ_{k+1}. If λ_k, λ_{k+1} are close, then the kth PC, $\boldsymbol{\alpha}_k'\mathbf{x} = z_k$, is more unstable than if λ_k, λ_{k+1} are well separated. A similar analysis can be carried out if λ_k is *increased* by an amount ε, in which case the stability of z_k depends on the separation between λ_k and λ_{k-1}. Thus, the stability of a PC depends on the separation of its variance from the variances of adjacent PCs, an unsurprising result, especially considering the discussion of 'influence' in Section 10.2. The ideas involved in Section 10.2, as here, are concerned with how perturbations effect $\boldsymbol{\alpha}_k$, but they differ in that the perturbations are deletions of individual observations, rather than hypothetical changes in λ_k. Nevertheless, we find in both cases that the changes in $\boldsymbol{\alpha}_k$ are largest if λ_k is close to λ_{k+1} or λ_{k-1}—see equation (10.2.4).

As an example of the use of the expression (12.4.1), consider again the PCs discussed earlier in this section, and originally presented in Table 3.2. It was found that rounding the coefficients in the PCs to the nearest 0.2 gave a change of 9% in λ_1, and changed the direction of $\boldsymbol{\alpha}_1$ through an angle of about 8°. We can use (12.4.1) to find the maximum angular change in $\boldsymbol{\alpha}_1$ which can occur if λ_1 is decreased by 9%. The maximum angle is, in fact, nearly 25°, so that rounding the coefficients has certainly not moved $\boldsymbol{\alpha}_1$ in the direction of maximum sensitivity.

λ_1, λ_2 and λ_3 in this example are 2.792, 1.532 and 1.250, respectively, so that the separation between λ_1 and λ_2 is much greater than that between λ_2

and λ_3. The potential change in $\boldsymbol{\alpha}_2$ for a given decrease in λ_2 will therefore be larger than that for $\boldsymbol{\alpha}_1$, given a corresponding decrease in λ_1. In fact, the same percentage decrease in λ_2 as that investigated above for λ_1 leads to a maximum angular change of 35°; if the change is made the same in absolute (rather than percentage) terms, then the maximum angle becomes 44°.

12.5. Principal Components in the Presence of Secondary or Instrumental Variables

Rao (1964) describes two modifications of PCA which involve what he calls 'instrumental variables'. These appear to be variables which are of secondary importance, but which may be useful in various ways in examining the variables which are of primary concern. The term 'instrumental variable' is in widespread use in econometrics, but in a rather more restricted context—see, for example, Harvey (1981, p. 78).

Suppose that $\mathbf{x}$ is, as usual, a p-element vector of primary variables, and that $\mathbf{w}$ is a vector of s secondary, or instrumental, variables. Rao (1964) considers the following two problems:

(i) find linear functions $\boldsymbol{\gamma}_1'\mathbf{w}, \boldsymbol{\gamma}_2'\mathbf{w}, \ldots$, of $\mathbf{w}$ which best predict $\mathbf{x}$;
(ii) find linear functions $\boldsymbol{\alpha}_1'\mathbf{x}, \boldsymbol{\alpha}_2'\mathbf{x}, \ldots$, with maximum variances which, as well as being uncorrelated with each other, are also uncorrelated with $\mathbf{w}$.

For (i), Rao (1964) notes that $\mathbf{w}$ may contain some or all of the elements of $\mathbf{x}$, and gives two possible measures of predictive ability, corresponding to the trace and Euclidean norm criteria discussed with respect to Property A5 in Section 2.1. He also mentions the possibility of introducing weights into the analysis. The two criteria lead to different solutions to (i), one of which is more straightforward to derive than the other. There is a superficial resemblance between the current problem and that of canonical correlation analysis, where relationships between two sets of variables are also investigated (see Section 9.3), but the two situations are easily seen to be different.

One situation mentioned by Rao (1964) where (ii) might be relevant is when the data $\mathbf{x}_1, \mathbf{x}_2, \ldots, \mathbf{x}_n$ form a multiple time series with p variables and n time points, and it is required to identify linear functions of $\mathbf{x}$ which have large variances, but which are independent of 'trend' in the time series—see Section 4.5 for an example where the first PC is dominated by trend. Rao argues that such functions can be found by defining instrumental variables which represent trend, and then solving the problem posed in (ii), but he gives no example to illustrate this idea.

Kloek and Mennes (1960) also discussed the use of PCs as 'instrumental variables', in an econometric context. In their analysis, a number of dependent variables, $\mathbf{y}$, are to be predicted from a set of regressor variables $\mathbf{x}$. Information is also available concerning another set of variables, $\mathbf{w}$ (the instrumental variables), which are not used directly in predicting $\mathbf{y}$, but which can be used to obtain improved estimates of the coefficients, **B**, in the equation predicting **y** from **x**. Kloek and Mennes (1960) examine a number of ways in which PCs of **w**, or PCs of the residuals obtained from regressing **w** on **x**, of PCs of the combined vector containing all elements of **w** *and* **x**, can be used as 'instrumental variables' in order to obtain improved estimates of the coefficients **B**.

12.6. Alternatives to Principal Component Analysis for Non-normal Distributions

We have noted several times that for many purposes it is not necessary to assume any particular distribution for the variables, **x**, in a PCA, although some of the properties of Chapters 2 and 3 rely on the assumption of multivariate normality. Section 10.3 discussed the idea of robustly estimating PCs in situations where some of the observations may be extreme, with respect to the distribution of the bulk of the data.

Another possibility is that the distribution of **x** is known to have a particular distribution, other than the multivariate normal. Again, for descriptive purposes a PCA can be very useful in such circumstances, but there are other alternatives, as suggested in unpublished work by O'Hagan at the University of Warwick. O'Hagan has noted that for a multivariate normal distribution several different properties characterize the PCs, but that the different properties will lead to different functions for non-normal distributions. In particular, the first PC is defined by

(a) the direction of maximum width of contours of equal probability (see Property G1 of Section 2.2);
(b) the direction of maximum width of the inflexion boundary for the distribution.

Also, the covariance matrix is given by the negative of the inverse of 'curvature' of the log probability density function, where 'curvature' is defined as the second derivative with respect to **x**. O'Hagan has suggested that each of these three properties could be used to define modifications of PCA for non-normal distributions, and has noted some problems associated with each. However, it appears that the ideas have not yet been explored in practice.

12.7. Three-Mode Principal Component Analysis

This short section gives a brief introduction to a topic which has recently been the subject of a 398-page book (Kroonenberg, 1983a); a 33-page annotated bibliography (Kroonenberg, 1983b) gives a comprehensive list of references for the slightly wider topic of three-mode factor analysis. The ideas for three-mode methods were first published by Tucker in the mid-1960s (see, for example, Tucker, 1966) and the term 'three-mode' refers to data sets which have three modes by which the data may be classified. For example, when PCs are obtained for several groups of individuals as in Section 11.5, there are three modes corresponding to variables, groups and individuals. Alternatively, we might have n individuals, p variables and t time points, so that 'individuals', 'variables' and 'time points' define the three modes. In this particular case we have effectively n time series of p variables, or a single time series of np variables. However, the time points need not be equally spaced, nor is the time-order of the t repetitions necessarily relevant in the sort of data for which three-mode PCA is used, in the same way that neither individuals or variables have any particular *a priori* ordering.

Let x_{ijk} be the observed value of the jth variable for the ith individual measured on the kth occasion. The basic idea in three-mode analysis is to approximate x_{ijk} by the model

$$\tilde{x}_{ijk} = \sum_{h=1}^{m} \sum_{l=1}^{q} \sum_{r=1}^{s} a_{ih} b_{jl} c_{kr} g_{hlr}.$$

The values m, q, s are less, and if possible very much less, than n, p, t respectively, and the parameters $a_{ih}, b_{jl}, c_{kr}, g_{hlr}, i = 1, 2, \ldots, n, h = 1, 2, \ldots, m, j = 1, 2, \ldots, p, l = 1, 2, \ldots, q, k = 1, 2, \ldots, t, r = 1, 2, \ldots, s$ are chosen to give a good fit of $\tilde{x}_{ijk}$ to x_{ijk} for all i, j, k. There are a number of methods for solving this problem and, like ordinary PCA, they involve finding eigenvalues and eigenvectors of cross-product or covariance matrices, in this case by combining two of the modes (e.g. combine individuals and observations to give a mode with np categories) before finding cross-products. Details will not be given here—see Tucker (1966) or Kroonenberg (1983a), where examples may also be found. Kroonenberg (1983b) indicates that three-mode PCA has been extended, in theory, to N-mode analysis ($N > 3$), but that this type of analysis has been rarely used in practice.

12.8. Concluding Remarks

It has been seen in this book, and particularly in the final two chapters, that PCA can be used in a wide variety of different ways. Many of the topics in the last two chapters are of recent origin, and it is likely that there will be further developments in the near future, which should help to clarify the usefulness, in practice, of some of the newer techniques. Further uses and adaptations of PCA are almost certain to be proposed and, given the large number of fields of application in which PCA has been employed, it seems highly probable that there are already some uses and modifications of which the present author is unaware. For example, Rao (1964) mentioned the possibility of modifying PCA if there are restrictions on the residual variances of the individual variables, after removing the first q components. It seems likely that, in the 20 years since Rao's paper was published, other work has been done on PCA under these, or related, constraints.

To conclude, we stress again that, far from being an old and narrow technique, PCA is the subject of much recent research and has great versatility, both in the ways in which it can be applied, and in the fields of application for which it is useful.

Computation of Principal Components

This Appendix consists of two sections, both concerned with the computation of PCs.

The first section describes efficient methods for calculating PCs, i.e. efficient techniques from numerical analysis for calculating eigenvectors, and eigenvalues, of positive semidefinite matrices.

The second section then discusses the facilities for computing PCs, and performing related analyses, which are available in five of the best known statistical computer packages.

A1. Numerical Calculation of Principal Components

Most users of PCA, whether statisticians or non-statisticians, will have little desire to know about efficient algorithms for computing PCs. Typically, a user will have access to a statistical program or package which performs the analysis automatically; relevant facilities which are available in some of the most widely-used packages are described briefly in the next section. Thus, the user does not need to write his or her own programs, and he or she will often have little or no interest in whether or not the package which is available performs its analysis efficiently. As long as the results emerge, the user is satisfied.

However, the type of algorithm used can be important, especially if some of the last few PCs, corresponding to small eigenvalues, are of interest. Many programs for PCA are geared to looking mainly at the first few PCs, especially if PCA is included only as part of a factor analysis routine; in this case, several algorithms can be used successfully, although some will encounter problems if any pair of the eigenvalues are very close together. When the last few, or all, of the PCs are to be calculated, problems can also arise for some algorithms, if some of the eigenvalues are very small.

Finding PCs reduces to finding the eigenvalues and eigenvectors of a positive-semidefinite matrix. We now look briefly at some of the possible algorithms which can be used to solve such an eigenvalue–eigenvector problem.

The Power Method

The only method of computation which has fairly widespread coverage in statistical textbooks (see, for example, Morrison, 1976, Section 8.4) is the power method. A form of the technique was, in fact, described by Hotelling (1933) in his original paper on PCA, and an accelerated version of the method was presented in Hotelling (1936). In its simplest form, the power method is a technique for finding the largest eigenvalue, and the corresponding eigenvector, of a $(p \times p)$ matrix $\mathbf{T}$. The idea is to choose an initial p-element vector $\mathbf{u}_0$, and then form the sequence

$$\mathbf{u}_1 = \mathbf{T}\mathbf{u}_0,$$

$$\mathbf{u}_2 = \mathbf{T}\mathbf{u}_1 \quad = \mathbf{T}^2\mathbf{u}_0,$$

$$\vdots \qquad \vdots \qquad \vdots$$

$$\mathbf{u}_r = \mathbf{T}\mathbf{u}_{r-1} = \mathbf{T}^r\mathbf{u}_0.$$

$$\vdots \qquad \vdots \qquad \vdots$$

If $\boldsymbol{\alpha}_1, \boldsymbol{\alpha}_2, \ldots, \boldsymbol{\alpha}_p$ are the eigenvectors of $\mathbf{T}$, then they form a basis for p-dimensional space, and we can write, for arbitrary $\mathbf{u}_0$,

$$\mathbf{u}_0 = \sum_{k=1}^{p} \kappa_k \boldsymbol{\alpha}_k$$

for some set of constants $\kappa_1, \kappa_2, \ldots, \kappa_p$. Then

$$\mathbf{u}_1 = \mathbf{T}\mathbf{u}_0 = \sum_{k=1}^{p} \kappa_k \mathbf{T}\boldsymbol{\alpha}_k = \sum_{k=1}^{p} \kappa_k \lambda_k \boldsymbol{\alpha}_k,$$

where $\lambda_1, \lambda_2, \ldots, \lambda_p$ are the eigenvalues of $\mathbf{T}$. Continuing, we get, for $r = 2, 3, \ldots,$

$$\mathbf{u}_r = \sum_{k=1}^{p} \kappa_k \lambda_k^r \boldsymbol{\alpha}_k,$$

and

$$\frac{\mathbf{u}_r}{(\kappa_1 \lambda_1^r)} = \left(\boldsymbol{\alpha}_1 + \frac{\kappa_2}{\kappa_1} \left(\frac{\lambda_2}{\lambda_1} \right)^r \boldsymbol{\alpha}_2 + \cdots + \frac{\kappa_p}{\kappa_1} \left(\frac{\lambda_p}{\lambda_1} \right)^r \boldsymbol{\alpha}_p \right).$$

Assuming that the first eigenvalue of $\mathbf{T}$ is distinct from the remaining eigenvalues, so that $\lambda_1 > \lambda_2 \geq \cdots \geq \lambda_p$, it follows that a suitably normalized version of $\mathbf{u}_r \to \boldsymbol{\alpha}_1$ as $r \to \infty$. It also follows that the ratios of corresponding elements of $\mathbf{u}_r$ and $\mathbf{u}_{r-1} \to \lambda_1$ as $r \to \infty$.

The power method thus gives a simple algorithm for finding the first (largest) eigenvalue of a covariance or correlation matrix, and its corresponding eigenvector, and hence the first PC and its variance. It works well if $\lambda_1 \gg \lambda_2$, but will converge only slowly if λ_1 is not much larger than λ_2. Speed of convergence will also depend on the choice of the initial vector $\mathbf{u}_0$; convergence will be most rapid if $\mathbf{u}_0$ is close to $\boldsymbol{\alpha}_1$.

If $\lambda_1 = \lambda_2 > \lambda_3$, a similar argument to that given above shows that a suitably normalized version of $\mathbf{u}_r \to \boldsymbol{\alpha}_1 + (\kappa_2/\kappa_1)\boldsymbol{\alpha}_2$ as $r \to \infty$. Thus, the method will still provide information about the space spanned by $\boldsymbol{\alpha}_1, \boldsymbol{\alpha}_2$. However, exact equality of eigenvalues is extremely unlikely for sample covariance or correlation matrices, so we need not really worry about this case.

Rather than looking at all $\mathbf{u}_r, r = 1, 2, 3, \ldots$, attention can be restricted to $\mathbf{u}_1, \mathbf{u}_2, \mathbf{u}_4, \mathbf{u}_8, \ldots$ (i.e. $\mathbf{T}\mathbf{u}_0, \mathbf{T}^2\mathbf{u}_0, \mathbf{T}^4\mathbf{u}_0, \mathbf{T}^8\mathbf{u}_0, \ldots$) by simply squaring each successive power of $\mathbf{T}$. This accelerated version of the power method was suggested by Hotelling (1936). The power method can be adapted to find the second, third, ... PCs, or the last few PCs (see Morrison, 1976, p. 281), but it is likely to encounter problems of convergence if eigenvalues are close together, and accuracy will diminish if several PCs are found by the method. Simple worked examples can be found in Hotelling (1936) and Morrison (1976, Section 8.4).

There are various adaptations to the power method which partially overcome some of the problems just mentioned. A large number of such adaptations are discussed by Wilkinson (1965, Chapter 9), although some are not directly relevant to positive-semidefinite matrices such as covariance or correlation matrices. Two ideas which are of use for such matrices will be mentioned here. First, the origin can be shifted, i.e. the matrix $\mathbf{T}$ is replaced by $\mathbf{T} - \rho\mathbf{I}_p$, where $\mathbf{I}_p$ is the identity matrix, and ρ is chosen to make the ratio of the first two eigenvalues of $\mathbf{T} - \rho\mathbf{I}_p$ much larger than the corresponding ratio for $\mathbf{T}$, hence speeding up convergence.

A second modification is to use inverse iteration (with shifts), in which case the iterations of the power method are used but with $(\mathbf{T} - \rho\mathbf{I}_p)^{-1}$ replacing $\mathbf{T}$. This modification has the advantage over the basic power method with shifts that, by using appropriate choices of ρ (different for different eigenvectors), convergence to *any* of the eigenvectors of $\mathbf{T}$ can be achieved. (For the basic method it is only possible to converge, in the first instance, to $\boldsymbol{\alpha}_1$ or to $\boldsymbol{\alpha}_p$.) Furthermore, it is not necessary to calculate explicitly the inverse of $\mathbf{T} - \rho\mathbf{I}_p$, because the equation $\mathbf{u}_r = (\mathbf{T} - \rho\mathbf{I}_p)^{-1}\mathbf{u}_{r-1}$ can be replaced by the equation $(\mathbf{T} - \rho\mathbf{I}_p)\mathbf{u}_r = \mathbf{u}_{r-1}$. The latter equation can then be solved using an efficient method for the solution of systems of linear equations (see Wilkinson, 1965,

Chapter 4). Overall, computational savings with inverse iteration can be
large compared to the basic power method (with or without shifts), especially
for matrices with special structure, such as tridiagonal matrices. It turns out
that an efficient way of computing PCs is to first transform the covariance or
correlation matrix to tridiagonal form using, for example, either the Givens
or Householder transformations (Wilkinson, 1965, pp. 282, 290), and then to
implement inverse iteration with shifts on this tridiagonal form.

   There is one problem with shifting the origin, which has not yet been
mentioned. This is the fact that to choose efficiently the values of ρ which
determine the shifts, we need some preliminary idea of the eigenvalues of $\mathbf{T}$.
This preliminary estimation can be achieved by using the method of bisection, which in turn is based on the Sturm sequence property of tridiagonal
matrices. Details will not be given here (see Wilkinson, 1965, pp. 300–302),
but the method provides a quick way of finding approximate values of the
eigenvalues of a tridiagonal matrix. In fact, bisection could be used to find the
eigenvalues to any required degree of accuracy, and inverse iteration used
solely to find the eigenvectors.

   There are two major collections of subroutines for finding eigenvalues and
eigenvectors for a wide variety of classes of matrix. These are the EISPACK
package (Smith et al., 1976), which is distributed by IMSL, and parts of the
NAG library of subroutines. In both of these collections, there are recom-
mendations as to which subroutines are most appropriate for various types
of eigenvalue–eigenvector problem. In the case where a few of the eigenvalues and eigenvectors of a real symmetric matrix are required (corresponding to finding just a few of the PCs for a covariance or correlation matrix)
both EISPACK and NAG recommend transforming to tridiagonal form,
using Householder transformations, and then finding eigenvalues and eigenvectors using bisection and inverse iteration respectively. NAG and
EISPACK both base their subroutines on algorithms published in Wilkinson
and Reinsch (1971).

The QL Algorithm

If all of the PCs are required, then methods other than those just described
may be more efficient. For example, both EISPACK and NAG recommend
that we should still transform the covariance or correlation matrix to tridiagonal form, but at the second stage the so-called QL algorithm should
now be used, instead of bisection and inverse iteration. Chapter 8 of Wilkinson (1965) spends over 80 pages describing the QR and LR algorithms (which
are closely related to the QL algorithm), but only a very brief outline will be
given here.

   The basic idea behind the QL algorithm is that any non-singular matrix $\mathbf{T}$
can be written as $\mathbf{T} = \mathbf{QL}$, where $\mathbf{Q}$ is orthogonal and $\mathbf{L}$ is lower triangular.

(The **QR** algorithm is similar, except that **T** is written instead as $\mathbf{T} = \mathbf{QR}$, where **R** is upper triangular, rather than lower triangular.) If $\mathbf{T}_1 = \mathbf{T}$, and we write $\mathbf{T}_1 = \mathbf{Q}_1\mathbf{L}_1$, then $\mathbf{T}_2$ is defined as $\mathbf{T}_2 = \mathbf{L}_1\mathbf{Q}_1$. This is the first step in an iterative procedure, in which $\mathbf{T}_r$ is written as $\mathbf{Q}_r\mathbf{L}_r$, and $\mathbf{T}_{r+1}$ is then defined as $\mathbf{L}_r\mathbf{Q}_r$, $r = 1, 2, 3, \ldots$, where $\mathbf{Q}_1, \mathbf{Q}_2, \mathbf{Q}_3, \ldots$, are orthogonal matrices, and $\mathbf{L}_1, \mathbf{L}_2, \mathbf{L}_3, \ldots$, are lower triangular. It can be shown that $\mathbf{T}_r$ converges to a diagonal matrix, with the eigenvalues of **T**, in decreasing absolute size, down the diagonal; eigenvectors can also be found.

As with the power method, the speed of convergence of the QL algorithm depends on the ratios of consecutive eigenvalues. The idea of incorporating shifts can again be implemented to improve the algorithm, and, unlike the power method, efficient strategies exist, for finding appropriate shifts, which do not rely on prior information about the eigenvalues—see, for example, Lawson and Hanson (1974, p. 109). The QL algorithm can also cope with equality between eigenvalues.

Singular Value Decomposition

It is probably fair to say that the algorithms described in detail by Wilkinson (1965) more than 20 years ago, and implemented in various EISPACK and NAG routines, have stood the test of time. Only minor improvements have been made. However, more recent sources which refer specifically to PCA, rather than general eigenvalue problems, suggest that PCs may best be computed using the singular value decomposition (SVD), which was discussed in Section 3.5. For example, Chambers (1977, p. 111) talks about the SVD '... providing the best approach to computation of principal components ...', and Gnanadesikan (1977, p. 10) states that '... the recommended algorithm for ... obtaining the principal components is either the ... QR method ... or the singular value decomposition ...'. In constructing the SVD, it turns out that similar algorithms to those given above can be used. For example, Lawson and Hanson (1974, p. 110) describe an algorithm—see also Wilkinson and Reinsch (1971)—for finding the SVD, which has two stages; the first uses Householder transformations to transform to an upper bidiagonal matrix, and the second applies an adapted QR algorithm. The method is therefore not radically different from that described earlier.

As noted at the end of Section 8.1, the SVD can also be useful in computations for regression (Mandel, 1982; Nelder, 1985), so the SVD has further advantages if PCA is used in conjunction with regression. Nash and Lefkovitch (1976) describe an algorithm which uses the SVD to provide a variety of results for regression, as well as PCs.

Another point concerning the SVD is that it provides simultaneously not only the coefficients and variances for the PCs, but also the scores of each observation on each PC, and hence the information which is required to construct a biplot (see Section 5.3). The PC scores would otherwise need to be derived *indirectly* from the eigenvalues and eigenvectors of the covariance or correlation matrix $\mathbf{S} = [1/(n-1)]\mathbf{X}'\mathbf{X}$.

The values of the PC scores are related to the eigenvectors of $\mathbf{XX}'$, which can be derived from the eigenvectors of $\mathbf{X}'\mathbf{X}$ (see the proof of Property G4 in Section 3.2); conversely, the eigenvectors of $\mathbf{X}'\mathbf{X}$ can be found from those of $\mathbf{XX}'$. In circumstances where the sample size (n) is smaller than the number of variables (p), $\mathbf{XX}'$ will have smaller dimensions than $\mathbf{X}'\mathbf{X}$, so that it can be advantageous to use the algorithms described above, based on the power method or QL method, on a multiple of $\mathbf{XX}'$, rather than $\mathbf{X}'\mathbf{X}$, in such cases.

A2. Principal Component Analysis in Computer Packages

This section of the Appendix provides brief descriptions of the facilities available for finding PCs, and also for related techniques such as PC regression (see Section 8.1), principal co-ordinates analysis (Section 5.2), biplots (Section 5.3), etc., within five widely-used computer packages.

The decision to keep the descriptions brief has been taken for two main reasons. The first is that the only satisfactory alternative is to have very lengthy descriptions; medium-length descriptions would not be adequate, because examples of the instructions needed to run the packages would only be useful if accompanying explanations were detailed enough to understand how to actually run the program. A considerable amount of space would be needed to achieve this. Inclusion of output from each program would again take up many pages, especially if it were all to be adequately explained. Furthermore, the output from PCA programs would be rather similar, though different in detail, for each of the packages examined.

A second, and probably more important, reason for brevity is that any description of computer packages inevitably becomes out-of-date quite rapidly, although the material below should give some idea of the minimum facilities available in each package. The expectation is that facilities will be added to, or improved, in the future, rather than reduced. Currently, most broad statistical packages will compute PCs, and will give various associated plots and statistics, but will not readily produce some of the more 'fringe' quantities, such as outlier detection statistics based on PCs (see Section 10.1), or biplots. However, some packages are very flexible and will allow the user to add extra pieces of programming to a package in order to carry out such additional analyses. These extra sections may be in a high-level language such as FORTRAN, or they may involve just a few simple instructions specific to the package. It seems likely that such flexibility will become a more widespread feature in the future. This flexibility will enable new techniques to be added (by users) to a package at an early stage of a technique's development. Techniques which subsequently become well established and widely used are then, at a later stage, likely to be incorporated into the main parts of the packages.

Two further developments which seem probable are the wider availability of good graphics, and easier interfacing between elements of different packages, but perhaps the most important development which is likely to occur in the near future is the increasing usage of personal, or micro, rather than mainframe, computers. Until recently, personal computers have not been powerful enough to handle the sort of computations needed to find PCs for large data sets. However, many packages, which include PCA, are now appearing for microcomputers, and already the 'Statistical Computing Software Review' section of the *American Statistician* is concentrating almost entirely on reviewing software for micro, rather than mainframe, computers.

Despite the likelihood that personal computers will become the main computing tool for an increasing number of users of PCA, the five statistical packages described below were developed for mainframe computers, and are discussed in that context (although some already have versions for microcomputers). At the time of writing, PCA is still usually carried out on mainframe computers, and the author has no experience yet of PCA on personal computers.

The proliferation of packages for PCA on microcomputers has a danger associated with it, namely that the algorithms used may not always be efficient, so that the user should be cautious in accepting at face value the results of any new package. This caveat is, of course, also valid for mainframe packages, although the user should be safe with well-tried packages such as those described below. For example, BMDP, GENSTAT and SAS* all use algorithms similar to those recommended in Section A1 above. BMDP, in fact, has adapted versions of the relevant EISPACK routines, GENSTAT uses Householder transformations followed by the QL algorithm, and SAS also uses the QL algorithm.

Five statistical packages, BMDP, GENSTAT, MINITAB, SAS, SPSS[x], are now discussed in a little more detail. Inevitably, some readers will feel that important packages have been omitted, but the five included here are certainly among the most widely used of the general statistical packages. One omission which should perhaps be noted here is S. This is rather different in nature from the five packages described below (it is perhaps closest in spirit to GENSTAT) in that it is much more a specialized *language* for data analysis than a *package* of predetermined sets of instructions for specific statistical methods. Still further along the road to languages is the language APL, which some people would argue is the most efficient vehicle for programming statistical techniques.

Among the other statistical packages which are available, some give a broad coverage of statistics, like the five to be discussed, while others concen-

---

* SAS is the registered trademark of SAS Institute Inc., Cary, NC, USA.

trate on particular areas within statistics; some are very specific in their coverage, e.g. PRINCALS (Gifi, 1983—see also Section 11.1), which concentrates entirely on PCA and some of its extensions. At the opposite end of the scale, large program libraries, such as NAG and IMSL, provide some statistical routines as one part of their very wide spectrum of facilities.

A2.1. BMDP

The BMDP suite of programs includes two which are relevant to PCA. The main one is P4M, which is a factor analysis program, with PCA as one option. To run a BMDP program, it is typically only necessary to enter a list of between 5 and 20 instructions, depending on the options which are desired. The analysis can be done for a correlation (default) or covariance matrix, and will provide simple plots of the values of the first few PCs (PC scores) two at a time. Plots of the coefficients in the PCs can also be made, and other output includes basic quantities such as the individual and cumulative variances of the PCs, as well as means, standard deviations and largest and smallest values of each of the original variables. The PC coefficients are given with the normalization $a_k' a_k = l_k$, rather than $a_k' a_k = 1$. A reason for sometimes preferring this type of normalization was discussed at the end of Section 2.3. Other options include finding non-centred PCs (see Section 12.3), and the possibility of assigning different weights to different observations (Section 12.1).

The program is, however, geared towards factor analysis, and hence towards the first few PCs. Because of the emphasis on factor analysis, the PCs are rotated unless a no-rotation option is selected, and PC scores are only available if this no-rotation option is chosen. In addition, only the PCs with an eigenvalue greater than unity (for correlation matrices) are retained automatically. The later PCs can be found by choosing a suitable option, but it is inconvenient to obtain plots of the last few PCs. This is because, at the time of writing, if a plot involving the jth and kth ($j < k$) PCs is required within P4M, then it is necessary to generate all $\frac{1}{2}k(k-1)$ plots involving the first k components.

Although the facilities within P4M are somewhat limited, it is possible to save statistics generated there, such as the PC scores, and then use them in another BMDP program. In this way it would be feasible, for example, to get plots with respect to the last few PCs, without automatically generating plots for all the earlier PCs as well, or to use the PC scores as variables in a further analysis, such as analysis of variance (see Section 11.4) or cluster analysis (see Section 9.2).

It is also possible to insert sections of FORTRAN code written by the user, and interfacing with other packages such as SAS (see Section A2.4 below) is available. Both of these features make the package more flexible than it might appear at first sight, albeit at the cost of additional effort.

The second BMDP program which involves PCA is P4R, which performs PC regression (see Section 8.1). Entry of the PCs (of the regressor variables) into the regression can depend on either the size of their variances, or on the size of the correlations between the dependent variable and the PCs.

Up-to-date information on BMDP can be obtained from

> BMDP Statistical Software Inc.,
> 1964 Westwood Boulevard, Suite 202,
> Los Angeles,
> California 90025,
> U.S.A.

A2.2. GENSTAT

Compared to many of the other packages, GENSTAT was written more for the statistician, although it is also used by non-statisticians. On first encounter, it is more difficult to use than other packages, but once a user is familiar with the language (a better word than 'package'—the manual refers to the GENSTAT language), GENSTAT provides a very powerful tool, especially for analyses which do not quite fit into a standard framework. A GENSTAT program consists of a list of *directives*, and the length of such a list will typically be a little shorter than the corresponding list of instructions for BMDP or SPSS^{X}.

The directive which performs PCA is called PCP. This directive will provide the first q (where q is chosen by the user) eigenvalues and eigenvectors of a covariance or correlation matrix, and hence the PCs. As well as coefficients (which have normalization $\mathbf{a}_k'\mathbf{a}_k = 1$), variances, and values for each of the first q PCs, PCP will give the residuals after fitting q PCs (which can be used to test for outliers—see Section 10.1), and a test for the equality of the last $p - q$ eigenvalues (see Section 3.7.3). Principal component scores, residuals, etc., can be stored for further analysis, such as plotting, which is done using separate directives. The PC scores can also be used as input to other analyses such as regression (see Section 8.1), analysis of variance (Section 11.4) or cluster analysis (Section 9.2).

In addition to the directive PCP, another relevant directive is PCO which implements principal co-ordinate analysis (see Section 5.2).

A powerful feature in GENSTAT is the opportunity for users to add facilities by means of *macros*. These are named sets of statements which can, essentially, be used as additional directives. It is relatively straightforward to construct GENSTAT macros for many of the techniques which are related to PCA. This is because most such techniques involve manipulation of matrices, and GENSTAT provides all the commonly used operations of matrix algebra as part of its general calculus.

There is also a standard library of macros which is periodically enlarged;

among those macros which are currently available are some which have relevance to PCA, such as BIPLOTV and CORRESP which respectively produce biplots (see Section 5.3) and implement correspondence analysis (Sections 5.4 and 11.1).

Up-to-date information regarding GENSTAT can be obtained from

> The GENSTAT Co-ordinator,
> Numerical Algorithms Group,
> 256 Banbury Road,
> Oxford, OX2 7DE,
> U.K.

A2.3. MINITAB

Unlike the other packages described in this Appendix, MINITAB does not have any direct instructions for finding PCs. It is included, however, because it is one of the most popular packages for teaching statistics, and so is often the first statistical package encountered by new users of statistics. Despite its lack of direct instructions, it is straightforward to find PCs in MINITAB, and, in fact, it is easier to construct biplots (see Section 5.3), or tests of outliers using PCs (see Section 10.1), for example, than in some of the more sophisticated packages. The reason for this flexibility is that MINITAB has simple instructions for handling matrices, including multiplication of matrices and evaluation of eigenvalues and eigenvectors. Principal component analysis can therefore be done by first using instructions for computing a correlation or covariance matrix, and then using the instruction for finding eigenvalues and eigenvectors of that matrix. The length of the list of MINITAB instructions for finding PCs will not greatly exceed that of other packages. Principal component scores, and also further statistics, such as biplots, can then be constructed using matrix multiplication. Additional analyses, for example PC regression (see Section 8.1), can be done on the PCs using other MINITAB instructions.

MINITAB is one of the easiest packages to use, especially interactively, but it is nevertheless very flexible, and can provide straightforward ways of performing quite sophisticated multivariate techniques. 'Macros' for carrying out multivariate, and other, methods are published from time to time in a MINITAB *Users' Group Newsletter*.

Up-to-date information on MINITAB can be obtained from

> MINITAB,
> 215 Pond Laboratory,
> University Park,
> Pennsylvania 16802,
> U.S.A.

A2.4. SAS

In the same way that BMDP consists of programs, and GENSTAT has directives, so SAS is made up of *procedures*. Principal component analysis can be done using one of two procedures, either FACTOR or PRINCOMP. As might be deduced from its name, FACTOR is really a factor analysis procedure, and PCA forms one special case within it. In most circumstances, PRINCOMP, which deals solely with PCA, will be preferred.

Because it does not need to include instructions concerning the various options within factor analysis, the list of instructions for PRINCOMP is generally shorter than the corresponding list for BMDP or SPSS^X. Also, the normalization $\mathbf{a}'_k\mathbf{a}_k = 1$ is used for the PC coefficients, unlike BMDP and SPSS^X which use $\mathbf{a}'_k\mathbf{a}_k = l_k$. PRINCOMP will produce as many PCs as are required, and give variances, coefficients and values for each of the PCs produced. Plots of PCs are obtained using the PLOT procedure, and it is no more difficult to get plots for the last few PCs than it is for the first few. Options include assigning different weights to different observations (see Section 12.1), doing separate PCAs for subsets of the observations, using a correlation or covariance matrix (rather than the matrix of observations) as input, or carrying out the PCA for *partial* covariances or correlations. Further analyses, such as cluster analysis (see Section 9.2) or regression analysis (Section 8.1), are possible, using other SAS procedures, on the results which are output from PRINCOMP. In particular, the MATRIX procedure is virtually a language itself and it allows sophisticated manipulation of matrices. Indeed, it could be used in place of the PRINCOMP procedure.

As mentioned earlier, interfacing is possible between SAS and BMDP.

Up-to-date information regarding SAS is available from a number of regional offices, including

SAS Institute Inc.,	and	SAS Software Ltd.
SAS Circle,		68 High Street,
PO Box 8000		Weybridge,
Cary,		Surrey, KT13 8BL,
North Carolina 27511,		U.K.
U.S.A.		

A2.5. SPSS^X

Like SAS, SPSS^X consists of *procedures*, but, unlike, SAS, there is no procedure specific to PCA alone. As in BMDP, PCA is available as one option within factor analysis, which is implemented for SPSS^X by the procedure FACTOR. The list of instructions necessary to run FACTOR is of similar length to that of P4M in BMDP, and, although the detailed options available in FACTOR differ in a number of ways from P4M, the two packages have many similarities when only PCA is required. For example, if PC scores, as well as the variances and coefficients of the PCs, are required, then it is necessary to invoke a no-rotation option. Also, the normalization $\mathbf{a}_k'\mathbf{a}_k = l_k$ is used for the PC coefficients, rather than the more usual $\mathbf{a}_k'\mathbf{a}_k = 1$. It is possible to save the PC scores, but, to output them, further PRINT or PLOT commands are apparently necessary, outside the FACTOR procedure. The saved scores could presumably also be used as input to other types of analysis, such as regression or cluster analysis (see Sections 8.1, 9.2 respectively). As in BMDP and SAS, different observations may be assigned different weights (see Section 12.1).

Principal components appear in one other part of SPSS^X, namely within the procedure MANOVA for multivariate analysis of variance (see Section 11.4). MANOVA includes an option which finds PCs of the error covariance or correlation matrix. Another feature of SPSS^X, which distinguishes it from the other packages discussed here, is that, as described in Section 11.6, it has various options for dealing with missing data. In the other packages, if an observation has values missing for any variables, then the observation is completely ignored.

Although the facilities for doing anything beyond a basic PCA in SPSS^X are perhaps more limited than for the other packages discussed here, the 1983 *User's Guide* describes some 'coming features' which may well be implemented by the time the present text is published. These include two additions which will be of particular use in increasing the flexibility of SPSS^X, and will enable it to go beyond a basic PCA. These features are a facility for users to write their own procedures, and a MATRIX procedure which appears to be similar in scope to that described above for SAS.

Up-to-date information on SPSS^X can be obtained from

SPSS Inc.,
444 N. Michigan Avenue,
Chicago,
Illinois 60611,
U.S.A.

References

AHAMAD, B. (1967). An analysis of crimes by the method of principal components. *Appl. Statist.*, **16**, 17–35.

AITCHISON, J. (1982). The statistical analysis of compositional data (with discussion). *J. R. Statist. Soc. B*, **44**, 139–177.

AITCHISON, J. (1983). Principal component analysis of compositional data. *Biometrika*, **70**, 57–65.

ALDENDERFER, M. S. and BLASHFIELD, R. K. (1984). *Cluster Analysis.* Beverly Hills: Sage.

ALLEN, D. M. (1974). The relationship between variable selection and data augmentation and a method for prediction. *Technometrics*, **16**, 125–127.

ANDERSON, J. R. and ROSEN, R. D. (1983). The latitude–height structure of 40–50 day variations in atmospheric angular momentum. *J. Atmos. Sci.*, **40**, 1584–1591.

ANDERSON, T. W. (1957). Maximum likelihood estimates for a multivariate normal distribution when some observations are missing. *J. Amer. Statist. Assoc.*, **52**, 200–203.

ANDERSON, T. W. (1963). Asymptotic theory for principal component analysis. *Ann. Math. Statist.*, **34**, 122–148.

ANDREWS, D. F. (1972). Plots of high-dimensional data. *Biometrics*, **28**, 125–136.

BARNETT, T. P. (1983). Interaction of the monsoon and Pacific trade wind system at interannual time scales. Part I: The equatorial zone. *Mon. Wea. Rev.*, **111**, 756–773.

BARNETT, V. (1981). *Interpreting Multivariate Data.* Chichester: Wiley.

BARNETT, V. and LEWIS, T. (1978). *Outliers in Statistical Data.* Chichester: Wiley.

BARTELS, C. P. A. (1977). *Economic Aspects of Regional Welfare, Income Distribution and Unemployment.* Leiden: Martinus Nijhoff.

BARTLETT, M. S. (1950). Tests of significance in factor analysis. *Brit. J. Psychol. Statist. Section*, **3**, 77–85.

BASKERVILLE, J. C. and TOOGOOD, J. H. (1982). Guided regression modeling for prediction and exploration of structure with many explanatory variables. *Technometrics*, **24**, 9–17.

BASSETT, E. E., CLEWER, A., GILBERT, P. and MORGAN, B. J. T. (1980). Forecasting numbers of households: the value of social and economic information. Unpublished report. University of Kent.

BEALE, E. M. L. and LITTLE, R. J. A. (1975). Missing values in multivariate analysis. *J. R. Statist. Soc. B*, **37**, 129–145.

BELSLEY, D. A. (1984). Demeaning conditioning diagnostics through centering (with comments). *Amer. Statistician*, **38**, 73–93.

BENZÉCRI, J.-P. (1980). *L'Analyse des Données*. Tome (Vol.)2: *L'Analyse des Correspondances*. Third edition. Paris: Dunod.

BERK, K. N. (1984). Validating regression procedures with new data. *Technometrics*, **26**, 331–338.

BESSE, P. and RAMSAY, J. O. (1984). Principal components analysis of interpolated functions. Paper presented at the 1984 Annual Meeting of the American Statistical Association.

BIBBY, J. (1980). Some effects of rounding optimal estimates. *Sankhya B*, **42**, 165–178.

BISHOP, Y. M. M., FIENBERG, S. E. and HOLLAND, P. W. (1975). *Discrete Multivariate Analysis: Theory and Practice*. Cambridge, MA: MIT Press.

BLACKITH, R. E. and REYMENT, R. A. (1971). *Multivariate Morphometrics*. London: Academic Press.

BLOOMFIELD, P. (1974). Linear transformations for multivariate binary data. *Biometrics*, **30**, 609–617.

BRILLINGER, D. R. (1981). *Time Series: Data Analysis and Theory*. Expanded edition. San Francisco: Holden-Day.

BRYANT, E. H. and ATCHLEY, W. R. (1975). *Multivariate Statistical Methods: Within-Group Covariation*. Stroudsberg: Halsted Press.

BUCKLAND, S. T. and ANDERSON, A. J. B. (1985). Multivariate analysis of Atlas data. In *Statistics in Ornithology* (eds. B. J. T. Morgan and P. M. North), 93–112. Berlin: Springer-Verlag.

CAHALAN, R. F. (1983). EOF spectral estimation in climate analysis. *Second International Meeting on Statistical Climatology*, Preprints Volume, 4.5.1–4.5.7.

CALDER, P. (1985). Influence functions for principal components and their variances. Submitted for publication.

CAMPBELL, N. A. (1980). Robust procedures in multivariate analysis I: Robust covariance estimation. *Appl. Statist.*, **29**, 231–237.

CATTELL, R. B. (1966). The scree test for the number of factors. *J. Multiv. Behav. Res.*, **1**, 245–276.

CATTELL, R. B. (1978). *The Scientific Use of Factor Analysis in Behavioral and Life Sciences*. New York: Plenum Press.

CHAMBERS, J. M. (1977). *Computational Methods for Data Analysis*. New York: Wiley.

CHAMBERS, J. M.. CLEVELAND, W. S., KLEINER, B. and TUKEY, P. A. (1983). *Graphical Methods for Data Analysis*. Belmont: Wadsworth.

CHANG, W.-C. (1983). On using principal components before separating a mixture of two multivariate normal distributions. *Appl. Statist.*, **32**, 267–275.

CHATFIELD, C. (1984). *The Analysis of Time Series: An Introduction*. Third edition. London: Chapman and Hall.

CHATFIELD, C. and COLLINS, A. J. (1980). *Introduction to Multivariate Analysis*. London: Chapman and Hall.

CHERNOFF, H. (1973). The use of faces to represent points in k-dimensional space graphically. *J. Amer. Statist. Assoc.*, **68**, 361–368.

CLEVELAND, W. S. and GUARINO, R. (1976). Some robust statistical procedures and their application to air pollution data. *Technometrics*, **18**, 401–409.

COCHRAN, R. N. and HORNE, F. H., (1977). Statistically weighted principal component analysis of rapid scanning wavelength kinetics experiments. *Anal. Chem.*, **49**, 846–853.

COLEMAN, D. (1985). Hotelling's T^2, robust principal components, and graphics for SPC. Paper presented at the 1985 Annual Meeting of the American Statistical Association.

COOK, R. D. and WEISBERG, S. (1982). *Residuals and Influence in Regression.* New York: Chapman and Hall.

CORSTEN, L. C. A. and GABRIEL, K. R. (1976). Graphical exploration in comparing variance matrices. *Biometrics,* **32,** 851–863.

COX, D. R. (1972). The analysis of multivariate binary data. *Appl. Statist.,* **21,** 113–120.

CRADDOCK, J. M. (1965). A meteorological application of principal component analysis. *Statistician,* **15,** 143–156.

CRADDOCK, J. M. and FLINTOFF, S. (1970). Eigenvector representations of Northern Hemispheric fields. *Q. J. R. Met. Soc.,* **96,** 124–129.

CRADDOCK, J. M. and FLOOD, C. R. (1969). Eigenvectors for representing the 500 mb. geopotential surface over the Northern Hemisphere. *Q. J. R. Met. Soc.,* **95,** 576–593.

CRITCHLEY, F. (1985). Influence in principal components analysis. *Biometrika.* **72,** 627–636.

DALING, J. R. and TAMURA, H. (1970). Use of orthogonal factors for selection of variables in a regression equation—an illustration. *Appl. Statist.,* **19,** 260–268.

DAVENPORT, M. and STUDDERT-KENNEDY, G. (1972). The statistical analysis of aesthetic judgment: an exploration. *Appl. Statist.,* **21,** 324–333.

DE LEEUW, J. (1982). Nonlinear principal component analysis. In *Compstat 82* (eds. H. Caussinus, P. Eltinger and R. Tomassone), 77–86. Wien: Physica-Verlag.

DE LEEUW, J. and VAN RIJCKEVORSEL, J. (1980). HOMALS and PRINCALS. Some generalizations of principal components analysis. In *Data Analysis and Informatics* (eds. E. Diday, L. Lebart, J. P. Pagès and R. Tomassone), 231–242. Amsterdam: North-Holland.

DE LIGNY, C. L., NIEUWDORP, G. H. E., BREDERODE, W. K., HAMMERS, W. E. and VAN HOUWELINGEN, J. C. (1981). An application of factor analysis with missing data. *Technometrics,* **23,** 91–95.

DENHAM, M. C. (1985). Unpublished postgraduate diploma project report. University of Kent.

DEVILLE, J.-C. and MALINVAUD, E. (1983). Data analysis in official socio-economic statistics (with discussion). *J. R. Statist. Soc. A,* **146,** 335–361.

DEVLIN, S. J., GNANADESIKAN, R. and KETTENRING, J. R. (1975). Robust estimation and outlier detection with correlation coefficients. *Biometrika,* **62,** 531–545.

DEVLIN, S. J., GNANADESIKAN, R. and KETTENRING, J. R. (1981). Robust estimation of dispersion matrices and principal components. *J. Amer. Statist. Assoc.,* **76,** 354–362.

DIACONIS, P. and EFRON, B. (1983). Computer-intensive methods in statistics. *Scientific Amer.,* **248,** 96–108.

DORAN, H. E. (1976). A spectral principal components estimator of the distributed lag model. *Int. Econ. Rev.,* **17,** 8–25.

DRAPER, N. R. and SMITH, H. (1981). *Applied Regression Analysis,* 2nd edition. New York: Wiley.

DURBIN, J. (1984). Time series analysis. Present position and potential developments: some personal views. *J. R. Statist. Soc. A,* **147,** 161–173.

DURBIN, J. and KNOTT, M. (1972). Components of Cramér–von Mises statistics. I. *J. R. Statist. Soc. B,* **34,** 290–307 (correction, **37,** 237).

DURBIN, J., KNOTT, M. and TAYLOR, C. C. (1975). Components of Cramér–von Mises statistics. II. *J. R. Statist. Soc. B,* **37,** 216–237.

EASTMENT, H. T. and KRZANOWSKI, W. J. (1982). Cross-validatory choice of the number of components from a principal component analysis. *Technometrics,* **24,** 73–77.

EVERITT, B. S. (1978). *Graphical Techniques for Multivariate Data*. London: Heinemann Educational Books.

EVERITT, B. S. (1980). *Cluster Analysis*, 2nd edition. London: Heinemann Educational Books.

FARMER, S. A. (1971). An investigation into the results of principal component analysis of data derived from random numbers. *Statistician*, **20**, 63–72.

FASHAM, M. J. R. (1977). A comparison of nonmetric multidimensional scaling, principal components and reciprocal averaging for the ordination of simulated coenoclines and coenoplanes. *Ecology*, **58**, 551–561.

FEDOROV, V. V. (1972). *Theory of Optimal Experiments*. New York: Academic Press.

FEENEY, G. J. and HESTER, D. D. (1967). Stock market indices: a principal components analysis. In *Risk Aversion and Portfolio Choice* (eds. D. D. Hester and J. Tobin), 110–138. New York: Wiley.

FELLEGI, I. P. (1975). Automatic editing and imputation of quantitative data. *Bull. Int. Statist. Inst.*, **46**, (3), 249–253.

FLURY, B. N. (1984). Common principal components in k groups. *J. Amer. Statist. Assoc.*, **79**, 892–898.

FLURY, B. N. (1985). Two generalizations of the common principal component model. Submitted for publication.

FLURY, B. and RIEDWYL, H. (1981). Graphical representation of multivariate data by means of asymmetrical faces. *J. Amer. Statist. Assoc.*, **76**, 757–765.

FOLLAND, C. K., PARKER, D. E. and NEWMAN, M. (1985). Worldwide marine temperature variations on the season to century time scale. *Proceedings of the Ninth Annual Climate Diagnostics Workshop*, 70–85.

FOMBY, T. B., HILL, R. C. and JOHNSON, S. R. (1978). An optimal property of principal components in the context of restricted least squares. *J. Amer. Statist. Assoc.*, **73**, 191–193.

FRANE, J. W. (1976). Some simple procedures for handling missing data in multivariate analysis. *Psychometrika*, **41**, 409–415.

FREEMAN, G. H. (1975). Analysis of interactions in incomplete two-way tables. *Appl. Statist.*, **24**, 46–55.

FRIEDMAN, S. and WEISBERG, H. F. (1981). Interpreting the first eigenvalue of a correlation matrix. *Educ. Psychol. Meas.*, **41**, 11–21.

FRISCH, R. (1929). Correlation and scatter in statistical variables. *Nordic Statist. J.*, **8**, 36–102.

GABRIEL, K. R. (1971). The biplot graphic display of matrices with application to principal component analysis. *Biometrika*, **58**, 453–467.

GABRIEL, K. R. (1978). Least squares approximation of matrices by additive and multiplicative models. *J. R. Statist. Soc. B*, **40**, 186–196.

GABRIEL, K. R. (1981). Biplot display of multivariate matrices for inspection of data and diagnosis. In *Interpreting Multivariate Data* (ed. V. Barnett), 147–173. Chichester: Wiley.

GABRIEL, K. R. and ODOROFF C. L. (1983). Resistant lower rank approximation of matrices. Technical report 83/02, Department of Statistics, University of Rochester.

GABRIEL, K. R. and ZAMIR, S. (1979). Lower rank approximation of matrices by least squares with any choice of weights. *Technometrics*, **21**, 489–498.

GARNHAM, N. (1979). Some aspects of the use of principal components in multiple regression. Unpublished M.Sc. dissertation, University of Kent.

GAUCH, H. G. (1982). Noise reduction by eigenvector ordinations. *Ecology*, **63**, 1643–1649.

GIFI, A. (1983). *PRINCALS User's Guide*. Leiden: University of Leiden.

GIRSHICK, M. A. (1936). Principal components. *J. Amer. Statist. Assoc.*, **31**, 519–528.

GIRSHICK, M. A. (1939). On the sampling theory of roots of determinantal equations. *Ann. Math. Statist.*, **10**, 203–224.

GITTINS, R. (1969). The application of ordination techniques. In *Ecological Aspects of the Mineral Nutrition of Plants* (ed. I. H. Rorison), 37–66. Oxford: Blackwell Scientific Publications.

GNANADESIKAN, R. (1977). *Methods for Statistical Data Analysis of Multivariate Observations*. New York: Wiley.

GNANADESIKAN, R. and KETTENRING, J. R. (1972). Robust estimates, residuals, and outlier detection with multiresponse data. *Biometrics*, **28**, 81–124.

GOLDSTEIN, M. and DILLON, W. R. (1978). *Discrete Discriminant Analysis*. New York: Wiley.

GORDON, A. D. (1981). *Classification. Methods for the Exploratory Analysis of Multivariate Data*. London: Chapman and Hall.

GOWER, J. C. (1966). Some distance properties of latent root and vector methods used in multivariate analysis. *Biometrika*, **53**, 325–338.

GOWER, J. C. (1967). Multivariate analysis and multidimensional geometry. *Statistician*, **17**, 13–28.

GREEN, B. F. (1977). Parameter sensitivity in multivariate methods. *J. Multiv. Behav. Res.*, **12**, 263–287.

GREENACRE, M. J. (1984). *Theory and Applications of Correspondence Analysis*. London: Academic Press.

GUIOT, J. (1981). Analyse mathématique de données geophysique. Applications à la dendroclimatologie. Unpublished Ph.D. dissertation, Université Catholique de Louvain.

GUNST, R. F. (1983). Regression analysis with multicollinear predictor variables: definition, detection and effects. *Commun. Statist.-Theor. Meth.*, **12**, 2217–2260.

GUNST, R. F. and MASON, R. L. (1977a). Biased estimation in regression: an evaluation using mean squared error. *J. Amer. Statist. Assoc.*, **72**, 616–628.

GUNST, R. F. and MASON, R. L. (1977b). Advantages of examining multicollinearities in regression analysis. *Biometrics*, **33**, 249–260.

GUNST, R. F. and MASON, R. L. (1980). *Regression Analysis and Its Applications: A Data-Oriented Approach*. New York: Dekker.

GUNST, R. F., WEBSTER, J. T. and MASON, R. L. (1976). A comparison of least squares and latent root regression estimators. *Technometrics*, **18**, 75–83.

HAMPEL, F. R. (1974). The influence curve and its role in robust estimation. *J. Amer. Statist. Assoc.*, **69**, 383–393.

HAND, D. J. (1982). *Kernel Discriminant Analysis*. Chichester: Research Studies Press.

HANSCH, C., LEO, A., UNGER, S. H., KIM, K. H., NIKAITANI, D. and LIEN, E. J. (1973). 'Aromatic' substituent constants for structure–activity correlations. *J. Medicinal Chem.*, **16**, 1207–1216.

HARTIGAN, J. A. (1975). *Clustering Algorithms*. New York: Wiley.

HARVEY, A. C. (1981). *The Econometric Analysis of Time Series*. Oxford: Philip Allan.

HAWKINS, D. M. (1973). On the investigation of alternative regressions by principal component analysis. *Appl. Statist.*, **22**, 275–286.

HAWKINS, D. M. (1974). The detection of errors in multivariate data using principal components. *J. Amer. Statist. Assoc.*, **69**, 340–344.

HAWKINS, D. M. (1980). *Identification of Outliers*. London: Chapman and Hall.

HAWKINS, D. M. and EPLETT, W. J. R. (1982). The Cholesky factorization of the inverse correlation or covariance matrix in multiple regression. *Technometrics*, **24**, 191–198.

HAWKINS, D. M. and FATTI, L. P. (1984). Exploring multivariate data using the minor principal components. *Statistician*, **33**, 325–338.

HILL, R. C., FOMBY, T. B. and JOHNSON, S. R. (1977). Component selection norms for principal components regression. *Commun. Statist.*, **A6**, 309–334.

HOAGLIN, D. C., MOSTELLER, F. and TUKEY, J. W. (1983). *Understanding Robust and

Exploratory Data Analysis. New York: Wiley.

HOCKING, R. R. (1976). The analysis and selection of variables in linear regression. *Biometrics*, **32**, 1–49.

HOCKING, R. R. (1984). Discussion of '*K*-clustering as a detection tool for influential subsets in regression' by J. B. Gray and R. F. Ling. *Technometrics*, **26**, 321–323.

HOCKING, R. R., SPEED, F. M. and LYNN, M. J. (1976). A class of biased estimators in linear regression. *Technometrics*, **18**, 425–437.

HOERL, A. E. and KENNARD, R. W. (1970a). Ridge regression: biased estimation for nonorthogonal problems. *Technometrics*, **12**, 55–67.

HOERL, A. E. and KENNARD, R. W. (1970b). Ridge regression: applications to nonorthogonal problems. *Technometrics*, **12**, 69–82.

HOERL, A. E., KENNARD, R. W. and HOERL, R. W. (1985). Practical use of ridge regression: a challenge met. *Appl. Statist.*, **34**, 114–120.

HOTELLING, H. (1933). Analysis of a complex of statistical variables into principal components. *J. Educ. Psychol.*, **24**, 417–441, 498–520.

HOTELLING, H. (1936). Simplified calculation of principal components. *Psychometrika*, **1**, 27–35.

HOTELLING, H. (1957). The relations of the newer multivariate statistical methods to factor analysis. *Brit. J. Statist. Psychol.*, **10**, 69–79.

HOUSEHOLDER, A. S. and YOUNG, G. (1938). Matrix approximation and latent roots. *Amer. Math. Mon.*, **45**, 165–171.

HSUAN, F. C. (1981). Ridge regression from principal component point of view. *Commun. Statist.*, **A10**, 1981–1995.

HUBER, P. J. (1981). *Robust Statistics.* New York: Wiley.

HUDLET, R. and JOHNSON, R. A. (1982). An extension of some optimal properties of principal components. *Ann. Inst. Statist. Math.*, **34**, 105–110.

HUNT, A. (1978). The elderly at home. *OPCS Social Survey Division*, Publication SS 1078. London: HMSO.

IGLARSH, H. J. and CHENG, D. C. (1980). Weighted estimators in regression with multicollinearity. *J. Statist. Computat. Simul.*, **10**, 103–112.

IMBER, V. (1977). A classification of the English personal social services authorities. *DHSS Statistical and Research Report Series.* No. 16. London: HMSO.

JACKSON, J. E. (1981). Principal components and factor analysis: Part III—What is factor analysis? *J. Qual. Tech.*, **13**, 125–130.

JACKSON, J. E. and HEARNE, F. T. (1973). Relationships among coefficients of vectors used in principal components. *Technometrics*, **15**, 601–610.

JACKSON, J. E. and HEARNE, F. T. (1979). Hotelling's T_M^2 for principal components— What about absolute values? *Technometrics*, **21**, 253–255.

JACKSON, J. E. and MUDHOLKAR, G. S. (1979). Control procedures for residuals associated with principal component analysis. *Technometrics*, **21**, 341–349.

JEFFERS, J. N. R. (1962). Principal component analysis of designed experiment. *Statistician*, **12**, 230–242.

JEFFERS, J. N. R. (1967). Two case studies in the application of principal component analysis. *Appl. Statist.*, **16**, 225–236.

JEFFERS, J. N. R. (1978). *An Introduction to Systems Analysis: With Ecological Applications.* London: Edward Arnold.

JEFFERS, J. N. R. (1981). Investigation of alternative regressions: some practical examples. *Statistician*, **30**, 79–88.

JOLLIFFE, I. T. (1970). Redundant variables in multivariate analysis. Unpublished D.Phil. thesis, University of Sussex.

JOLLIFFE, I. T. (1972). Discarding variables in a principal component analysis, I: Artificial data. *Appl. Statist.*, **21**, 160–173.

JOLLIFFE, I. T. (1973). Discarding variables in a principal component analysis, II: Real data. *Appl. Statist.*, **22**, 21–31.

JOLLIFFE, I. T. (1982). A note on the use of principal components in regression.

Appl. Statist., **31**, 300–303.

JOLLIFFE, I. T., JONES, B. and MORGAN, B. J. T. (1980). Cluster analysis of the elderly at home: a case study. In *Data Analysis and Informatics* (eds. E. Diday, L. Lebart, J. P. Pagès and R. Tomassone), 745–757. Amsterdam: North-Holland.

JOLLIFFE, I. T., JONES, B. and MORGAN, B. J. T. (1982a). An approach to assessing the needs of the elderly. *Clearing House for Local Authority Social Services Research*, **2**, 1–102.

JOLLIFFE, I. T., JONES, B. and MORGAN, B. J. T. (1982b). Utilising clusters: a case-study involving the elderly. *J. R. Statist. Soc. A*, **145**, 224–236.

JOLLIFFE, I. T., JONES, B. and MORGAN, B. J. T. (1986). Comparison of cluster ana-lyses of the English personal social services authorities. To appear.

JONES, P. D., WIGLEY, T. M. L. and BRIFFA, K. R. (1983). Reconstructing surface pressure patterns using principal components regression on temperature and pre-cipitation data. *Second International Meeting on Statistical Climatology*, Preprints Volume, 4.2.1–4.2.8.

JÖRESKOG, K. G., KLOVAN, J. E. and REYMENT, R. A. (1976). *Geological Factor Analysis*. Amsterdam: Elsevier.

KAISER, H. F. (1960). The application of electronic computers to factor analysis. *Educ. Psychol. Meas.*, **20**, 141–151.

KENDALL, M. G. (1957). *A Course in Multivariate Analysis*. London: Griffin.

KENDALL, M. G. (1966). Discrimination and classification. In *Multivariate Analysis* (ed. P. R. Krishnaiah), 165–185. New York: Academic Press.

KENDALL, M. G. and STUART, A. (1979). *The Advanced Theory of Statistics*, Vol. 2, 4th edition. London: Griffin.

KLOEK, T. and MENNES, L. B. M. (1960). Simultaneous equations estimation based on principal components of predetermined variables. *Econometrica*, **28**, 45–61.

KROONENBERG, P. M. (1983a). *Three-Mode Principal Component Analysis*. Leiden: DSWO Press.

KROONENBERG, P. M. (1983b). Annotated bibliography of three-mode factor analy-sis. *Brit. J. Math. Statist. Psychol.*, **36**, 81–113.

KRUSKAL, J. B. (1964a). Multidimensional scaling by optimizing goodness of fit to a nonmetric hypothesis. *Psychometrika*, **29**, 1–27.

KRUSKAL, J. B. (1964b). Nonmetric multidimensional scaling: a numerical method. *Psychometrika*, **29**, 115–129.

KRZANOWSKI, W. J. (1979a). Some exact percentage points of a statistic useful in analysis of variance and principal component analysis. *Technometrics*, **21**, 261–263.

KRZANOWSKI, W. J. (1979b). Between-groups comparison of principal components. *J. Amer. Statist. Assoc.*, **74**, 703–707 (correction **76**, 1022).

KRZANOWSKI, W. J. (1982). Between-group comparison of principal components—some sampling results. *J. Statist. Computat. Simul.*, **15**, 141–154.

KRZANOWSKI, W. J. (1983). Cross-validatory choice in principal component analysis: some sampling results. *J. Statist. Computat. Simul.*, **18**, 299–314.

KRZANOWSKI, W. J. (1984a). Principal component analysis in the presence of group structure. *Appl. Statist.*, **33**, 164–168.

KRZANOWSKI, W. J. (1984b). Sensitivity of principal components. *J. R. Statist. Soc. B*, **46**, 558–563.

KUNG, E. C. and SHARIF, T. A. (1980). Multi-regression forecasting of the Indian summer monsoon with antecedent patterns of the large-scale circulation. *WMO Symposium on Probabilistic and Statistical Methods in Weather Forecasting*, 295–302.

LACHENBRUCH, P. A. (1975). *Discriminant Analysis*. New York: Hafner Press.

LAWLEY, D. N. (1963). On testing a set of correlation coefficients for equality. *Ann. Math. Statist.*, **34**, 149–151.

LAWLEY, D. N. and MAXWELL, A. E. (1971). *Factor Analysis as a Statistical Method*,

2nd edition. London: Butterworth.

LAWSON, C. L. and HANSON, R. J. (1974). *Solving Least Squares Problems*. Englewood Cliffs, NJ: Prentice-Hall.

LEAMER, E. E. and CHAMBERLAIN, G. (1976). A Bayesian interpretation of pretesting. *J. R. Statist. Soc. B*, **38**, 85–94.

LEGENDRE, L. and LEGENDRE, P. (1983). *Numerical Ecology*. Amsterdam: Elsevier.

LI, G. and CHEN, Z. (1985). Projection-pursuit approach to robust dispersion matrices and principal components: primary theory and Monte Carlo. *J. Amer. Statist. Assoc.*, **80**, 759–766.

LOTT, W. F. (1973). The optimal set of principal component restrictions on a least-squares regression. *Commun. Statist.*, **2**, 449–464.

MACDONELL, W. R. (1902). On criminal anthropometry and the identification of criminals. *Biometrika*, **1**, 177–227.

MAGER, P. P. (1980a). Principal component regression analysis applied in structure–activity relationships. 2. Flexible opioids with unusually high safety margin. *Biom. J.*, **22**, 535–543.

MAGER, P. P. (1980b). Correlation between qualitatively distributed predicting variables and chemical terms in acridine derivatives using principal component analysis. *Biom. J.*, **22**, 813–825.

MANDEL, J. (1971). A new analysis of variance model for non-additive data. *Technometrics*, **13**, 1–18.

MANDEL, J. (1972). Principal components, analysis of variance and data structure. *Statistica Neerlandica*, **26**, 119–129.

MANDEL, J. (1982). Use of the singular value decomposition in regression analysis. *Amer. Statistician*, **36**, 15–24.

MANSFIELD, E. R., WEBSTER, J. T. and GUNST, R. F. (1977). An analytic variable selection technique for principal component regression. *Appl. Statist.*, **26**, 34–40.

MARDIA, K. V., KENT, J. T. and BIBBY, J. M. (1979). *Multivariate Analysis*. London: Academic Press.

MARQUARDT, D. W. (1970). Generalized inverses, ridge regression, biased linear estimation, and nonlinear estimation. *Technometrics*, **12**, 591–612.

MARYON, R. H. (1979). Eigenanalysis of the Northern Hemispherical 15-day mean surface pressure field and its application to long-range forecasting. *Met O 13 Branch Memorandum No. 82* (unpublished). UK Meteorological Office, Bracknell.

MASSY, W. F. (1965). Principal components regression in exploratory statistical research. *J. Amer. Statist. Assoc.*, **60**, 234–256.

MATTHEWS, J. N. S. (1984). Robust methods in the assessment of multivariate normality. *Appl. Statist.*, **33**, 272–277.

MAXWELL, A. E. (1977). *Multivariate Analysis in Behavioural Research*. London: Chapman and Hall.

McCABE, G. P. (1982). Principal variables. *Technical Report No. 82–3*, Department of Statistics, Purdue University.

McCABE, G. P. (1984). Principal variables. *Technometrics*, **26**, 137–144.

McREYNOLDS, W. O. (1970). Characterization of some liquid phases. *J. Chromatogr. Sci.*, **8**, 685–691.

MILLER, A. J. (1984). Selection of subsets of regression variables (with discussion). *J. R. Statist. Soc. A*, **147**, 389–425.

MORGAN, B. J. T. (1981). Aspects of QSAR: I. Unpublished report, CSIRO Division of Mathematics and Statistics, Melbourne.

MORRISON, D. F. (1976). *Multivariate Statistical Methods*, 2nd edition. Tokyo: McGraw-Hill Kogakusha.

MOSER, C. A. and SCOTT, W. (1961). *British Towns*. Edinburgh: Oliver and Boyd.

MOSTELLER, F. and TUKEY, J. W. (1977). *Data Analysis and Regression: A Second Course in Statistics*. Reading, MA: Addison-Wesley.

MULLER, K. E. (1982). Understanding canonical correlation through the general linear model and principal components. *Amer. Statistician*, **36**, 342–354.

NAES, T. (1985). Multivariate calibration when the error covariance matrix is structured. *Technometrics*, **27**, 301–311.

NASH, J. C. and LEFKOVITCH, L. P. (1976). Principal components and regression by singular value decomposition on a small computer. *Appl. Statist.*, **25**, 210–216.

NELDER, J. A. (1985). An alternative interpretation of the singular-value decomposition in regression. *Amer. Statistician*, **39**, 63–64.

O'HAGAN, A. (1984). Motivating principal components, and a stronger optimality result. *Statistician*, **33**, 313–315.

OKAMOTO, M. (1969). Optimality of principal components. In *Multivariate Analysis II* (ed. P. R. Krishnaiah), 673–685. New York: Academic Press.

OMAN, S. D. (1978). A Bayesian comparison of some estimators used in linear regression with multicollinear data. *Commun. Statist.* **A7**, 517–534.

OSMOND, C. (1985). Biplot models applied to cancer mortality rates. *Appl. Statist.*, **34**, 63–70.

OVERLAND, J. E. and PREISENDORFER, R. W. (1982). A significance test for principal components applied to a cyclone climatology. *Mon. Wea. Rev.*, **110**, 1–4.

PEARCE, S. C. and HOLLAND, D. A. (1960). Some applications of multivariate methods in botany. *Appl. Statist.*, **9**, 1–7.

PEARSON, K. (1901). On lines and planes of closest fit to systems of points in space. *Phil. Mag.* (6), **2**, 559–572.

PREISENDORFER, R. W. (1981). Principal component analysis and applications. Unpublished lecture notes. Amer. Met. Soc. Workshop on Principal Component Analysis, Monterey.

PREISENDORFER, R. W. and MOBLEY, C. D. (1982). Data intercomparison theory, I–V. NOAA Tech. Memoranda ERL PMEL Nos. 38–42.

PRESS, S. J. (1972). *Applied Multivariate Analysis*. New York: Holt, Rinehart and Winston.

PRIESTLEY, M. B. (1981). *Spectral Analysis and Time Series*, Volumes 1 and 2. London: Academic Press.

PRIESTLEY, M. B., SUBBA RAO, T. and TONG, H. (1974). Applications of principal component analysis and factor analysis in the identification of multivariable systems. *IEEE Trans. Automat. Contr.*, **AC-19**, 730–734.

RADHAKRISHNAN, R. and KSHIRSAGAR, A. M. (1981). Influence functions for certain parameters in multivariate analysis. *Commun. Statist.*, **A10**, 515–529.

RAO, C. R. (1955). Estimation and tests of significance in factor analysis. *Psychometrika*, **20**, 93–111.

RAO, C. R. (1958). Some statistical methods for comparison of growth curves. *Biometrics*, **14**, 1–17.

RAO, C. R. (1964). The use and interpretation of principal component analysis in applied research. *Sankhya A*, **26**, 329–358.

RAO, C. R. (1973). *Linear Statistical Inference and Its Applications*. 2nd edition. New York: Wiley.

RASMUSSON, E. M., ARKIN, P. A., CHEN, W-Y. and JALICKEE, J. B. (1981). Biennial variations in surface temperature over the United States as revealed by singular decomposition. *Mon. Wea. Rev.*, **109**, 587–598.

RAVEH, A. (1985). On the use of the inverse of the correlation matrix in multivariate data analysis. *Amer. Statistician*, **39**, 39–42.

RICHMAN, M. B. (1983). Specification of complex modes of circulation with *T*-mode factor analysis. *Second International Meeting on Statistical Climatology*, Preprints Volume, 5.1.1–5.1.8.

ROBERT, P. and ESCOUFIER, Y. (1976). A unifying tool for linear multivariate statistical methods: the RV coefficient. *Appl. Statist.*, **25**, 257–265.

RUMMEL, R. J. (1970). *Applied Factor Analysis*. Evanston: Northwestern University Press.

RUYMGAART, F. H. (1981). A robust principal component analysis. *J. Multiv. Anal.*, **11**, 485–497.

SCLOVE, S. L. (1968). Improved estimators for coefficients in linear regression. *J. Amer. Statist. Assoc.*, **63**, 596–606.

SIBSON, R. (1984). Multivariate analysis. Present position and potential developments: some personal views. *J. R. Statist. Soc. A*, **147**, 198–207.

SILVEY, S. D. (1980). *Optimal Design: An Introduction to the Theory for Parameter Estimation*. London: Chapman and Hall.

SMITH, B. T., BOYLE, J. M., DONGARRA, J. J., GARBOW, B. S., IKEBE, Y., KLEMA, V. C., and MOLER, C. B. (1976). *Matrix Eigensystem Routines—EISPACK guide*, 2nd edition. Berlin: Springer-Verlag.

SPRENT, P. (1972). The mathematics of size and shape. *Biometrics*, **28**, 23–37.

SPURRELL, D. J. (1963). Some metallurgical applications of principal components. *Appl. Statist.*, **12**, 180–188.

SRIVASTAVA, M. S. and KHATRI, C. G. (1979). *An Introduction to Multivariate Statistics*. New York: North Holland.

STEIN, C. M. (1960). Multiple regression. In *Contributions to Probability and Statistics. Essays in Honour of Harold Hotelling*, (ed. I. Olkin), 424–443. Stanford: Stanford University Press.

STONE, E. A. (1984). Cluster analysis of English counties according to socio-economic factors. Unpublished undergraduate dissertation. University of Kent.

STONE, R. (1947). On the interdependence of blocks of transactions (with discussion). *J. R. Statist. Soc. B*, **9**, 1–45.

STUART, M. (1982). A geometric approach to principal components analysis. *Amer. Statistician*, **36**, 365–367.

SUGIYAMA, T. and TONG, H. (1976). On a statistic useful in dimensionality reduction in multivariable linear stochastic system. *Commun. Statist.*, **A5**, 711–721.

SYLVESTRE, E. A., LAWTON, W. H. and MAGGIO, M. S. (1974). Curve resolution using a postulated chemical reaction. *Technometrics*, **16**, 353–368.

TER BRAAK, C. J. F. (1983). Principal components biplots and alpha and beta diversity. *Ecology*, **64**, 454–462.

THURSTONE, L. L. (1931). Multiple factor analysis. *Psychol. Rev.*, **38**, 406–427.

TORGERSON, W. S. (1958). *Theory and Methods of Scaling*. New York: Wiley.

TORTORA, R. D. (1980). The effect of a disproportionate stratified design on principal component analysis used for variable elimination. *Proceedings of the Amer. Statist. Assoc.* Section on Survey Research Methods, 746–750.

TRENKLER, G. (1980). Generalized mean squared error comparisons of biased regression estimators. *Commun. Statist.*, **A9**, 1247–1259.

TRYON, R. C. (1939). *Cluster Analysis*. Ann Arbor: Edwards Brothers.

TUCKER, L. R. (1966). Some mathematical notes on three-mode factor analysis. *Psychometrika*, **31**, 279–311.

TUKEY, P. A. and TUKEY, J. W. (1981). Graphical display of data sets in three or more dimensions. Three papers in *Interpreting Multivariate Data* (ed. V. Barnett), 189–275. Chichester: Wiley.

VELICER, W. F. (1976). Determining the number of components from the matrix of partial correlations. *Psychometrika*, **41**, 321–327.

WALKER, M. A. (1967). Some critical comments on 'An analysis of crimes by the method of principal components' by B. Ahamad. *Appl. Statist.*, **16**, 36–38.

WALLACE, T. D. (1972). Weaker criteria and tests for linear restrictions in regression. *Econometrica*, **40**, 689–698.

WANG, P. C. C. (1978). *Graphical Representation of Multivariate Data*. New York: Academic Press.

WEBBER, R. and CRAIG, J. (1978). Socio-economic classification of local authority areas. *OPCS Studies on Medical and Population Subjects*, No. 35. London: HMSO.

WEBSTER, J. T., GUNST, R. F. and MASON, R. L. (1974). Latent root regression analysis. *Technometrics*, **16**, 513–522.

WHITE, J. W. and GUNST, R. F. (1979). Latent root regression: large sample analysis. *Technometrics*, **21**, 481–488.

WIBERG, T. (1976). Computation of principal components when data are missing. In *Compstat 1976* (eds. J. Gordesch and P. Naeve), 229–236. Wien: Physica-Verlag.

WIGLEY, T. M. L., LOUGH, J. M. and JONES, P. D. (1984). Spatial patterns of precipitation in England and Wales and a revised, homogeneous England and Wales precipitation series. *J. Climatol.*, **4**, 1–25.

WILKINSON, J. H. (1965). *The Algebraic Eigenvalue Problem*. Oxford: Oxford University Press.

WILKINSON, J. H. and REINSCH, C. (1971). *Handbook for Automatic Computation*. Vol. II, *Linear Algebra*. Berlin: Springer-Verlag.

WOLD, S. (1976). Pattern recognition by means of disjoint principal components models. *Patt. Recog.*, **8**, 127–139.

WOLD, S. (1978). Cross-validatory estimation of the number of components in factor and principal components models. *Technometrics*, **20**, 397–405.

WOLD, S., ALBANO, C., DUNN, W. J., ESBENSEN, K., HELLBERG, S., JOHANSSON, E. and SJÖSTRÖM, M. (1983). Pattern recognition: finding and using regularities in multivariate data. In *Food Research and Data Analysis* (eds. H. Martens and H. Russwurm), 147–188. London: Applied Science Publishers.

WORTON, B. J. (1984). Statistical aspects of data collected in year 1974–1975 of the Irish Wetlands enquiry. Unpublished M.Sc. dissertation. University of Kent.

YULE, W., BERGER, M., BUTLER, S., NEWHAM, V. and TIZARD, J. (1969). The WPPSI: an empirical evaluation with a British sample. *Brit. J. Educ. Psychol.*, **39**, 1–13.

Index